Pesticides in Drinking Water

Pesticides in Drinking Water

David I. Gustafson, Ph.D.
Chapel Hill, North Carolina, USA

VNR VAN NOSTRAND REINHOLD
_____ New York

Printed in the United States of America
For more information, contact:

Van Nostrand Reinhold
115 Fifth Avenue
New York, NY 10003

International Thomson Publishing GmbH
Königswinterer Strasse 418
53227 Bonn
Germany

International Thomson Publishing Europe
Berkshire House 168–173
High Holborn
London WC1V 7AA
England

International Thomson Publishing Asia
221 Henderson Road #05 10
Henderson Building
Singapore 0315

Thomas Nelson Australia
102 Dodds Street
South Melbourne, 3205
Victoria, Australia

International Thomson Publishing Japan
Hirakawacho Kyowa Building, 3F
2-2-1 Hirakawacho
Chiyoda-ku, 102 Tokyo
Japan

Nelson Canada
1120 Birchmount Road
Scarborough, Ontario
Canada M1K 5G4

International Thomson Editores
Campos Eliseos 385, Piso 7
Col. Polanco
11560 Mexico D.F. Mexico

2 3 4 5 6 7 8 9 10 BBR 99 98 97 96 95

Library of Congress Cataloging-in-Publication Data

Gustafson, David I.
    Pesticides in drinking water / David I. Gustafson.
        p.   cm.
    Includes index.
    ISBN 0-442-01187-3
    1. Pesticides—Environmental aspects.   2. Drinking water—
Contamination.   3. Water quality management.   I. Title.
TD427.P35G87   1993
628.1'6842—dc20
                                                    92-33462
                                                        CIP

To Sima, Mauri, and Drew

Contents

Preface

The task of writing this book represented a unique opportunity to assemble, within one package, the host of observational, theoretical, and regulatory activities surrounding the issue of pesticides in drinking water. For those who have not delved deeply into the subject, this book provides extensive background information about past incidents, the routes by which these chemicals get into drinking water, and what's being done about it. For those who are now or have been more intimately involved, the text should still provide useful data because of its in-depth coverage of the various historical, technical, regulatory, and political aspects of pesticides in drinking water.

As water supplies are stretched thinner and our economically driven use of pesticides on farm land becomes even more intensive, the strains on water quality will become ever greater. It is by no means certain, however, that drinking water is fated to contain pesticides. As pointed out in the final chapters of this book, much can and is being done to control the unwanted off-site movement of these materials, and the vast majority of our drinking water today is thought to contain only negligible quantities of such chemicals.

A semantic note: The word "contamination" has been the source of some contentious debate between environmental activists and those more forgiving of chemicals in our midst. In its most narrow sense, contamination refers to the presence of chemicals or other materials at levels high enough to render the water undrinkable. The broadest use of the term contamination refers to the presence of any detectable amount of chemical in the water being analyzed. This text will lean toward the narrower use of the term, because this mirrors the way that the pesticide residues in drinking water are being managed: in a differential manner, depending on the concentrations found. Certain levels are regarded as safe and not requiring any remediation. If the broader use of the term were used, then it would be logical to assume that all water would eventually become contaminated, as our analytical abilities to detect successively smaller quantities grow more precise.

As my interest and work on the issue of pesticides in drinking water has moved forward, there have been many colleagues who have contributed to my thinking. While working at Shell Development Company in its former Modesto location, my mentor was Paul Porter, who introduced me to the theories governing pesticide transport through soil. At Monsanto in the mid- to late 1980s, I was privileged to work with many talented and inspiring scientists, including Larry Holden, Jeff Graham, and Drew Klein. More recently, I have had the opportunity to work with Russell Jones, whose pioneering experimental approaches to describing pesticide transport through soil have impressed and influenced me greatly. To each I extend my thanks and respectfully acknowledge their contributions to the many concepts summarized in this book.

Finally, I'd like to extend a special thanks to two scientists, Don Wauchope and Stuart Cohen, who—through their careful and thoughtful review of the draft manuscript—have helped bring overall balance and technical clarity to the final product.

I

How Big is the Problem?

1

Case Studies of Contamination Episodes

In a never-ending struggle to survive, man must eat. This need spawned the development and spread of agriculture across the planet in several waves, one of which is now thought to be responsible for the spread of Indo-European languages across Europe beginning some 10,000 years ago. As agricultural methods were refined and societies began to flourish, man became not just a feature and pawn of the environment, but was able to alter and mold it in both useful and, more recently, potentially harmful ways. One need only briefly peruse the recent pictures and reports coming from the recently environmentally negligent countries of Eastern Europe to see what can happen when technology runs amuck and does not heed the needs of the Earth.

As farming techniques have grown successively more sophisticated, the chemical industry has fed the continuing spiral of the ever-increasing quantity and quality of agricultural products by supplying growers with the chemical tools we now call pesticides. These chemical products eliminate or suppress various pests—from both the plant and animal kingdoms—allowing higher yields and higher-quality products to be supplied to society.

The science of pesticide use continues to evolve in different ways in different parts of the world. Taking as an example herbicide use in the midwestern United States, there have been four stages characterized by method of application (Pike, McGlamery, and Knake 1991). The first stage involved foliar sprays on the leaf surfaces of weeds after they had emerged from the soil—the so-called postemergent sprays. The second stage of herbicide development was marked by the use of prophylactic preemergent applications of herbicides directly to the soil surface. In the third stage, more selective materials were found that could be incorporated into the soil prior to planting and yet not hamper crop germination and emergence. These first

3

three stages were marked by an increased reliance upon consistent use of chemicals as well as increased levels of soil disturbance. The current stage is typified by slowly falling pesticide usage and dramatic reductions in soil tillage operations.

The beneficial effects of using the chemical tools (our bountiful supply of blemish-free fruits and vegetables is just one example) are now widely taken for granted and have given rise to very high consumer expectations that can be met only through the continued use of such chemicals. Unfortunately, these same chemical tools strain the abilities of the ecosystem in which they are placed to fully degrade them before they move off-site into the water and atmosphere where their toxic properties are unwanted and—in a very few cases—potentially dangerous to humans or other so-called "nontarget" species.

In this book, the term pesticides will be used generically in reference to all chemical products used by farmers to control insects, weeds, fungi, and other vectors of disease. As such, the term includes virtually all chemical products used in agriculture, with the exception of fertilizers. Another category of chemical products—plant growth regulators—is generally regulated in the same manner as pesticides, but none of these products will receive any further attention in this book.

In general, those pesticides meant to control insects (insecticides) are more acutely toxic than other materials are to nontarget species such as man, and it is generally insecticides that are most often thought of by the lay public when the word pesticide is mentioned. Considerably greater than insecticides in usage, however, are herbicides. These products, which can be either selective or nonselective, kill plants. If they possess sufficient selectivity, they may be placed directly onto the crop or into the soil where the crop is grown. If they are capable of harming the crop, then they are applied as a directed spray away from the crop leaves and roots or are used as a spot treatment. Nonselective herbicides such as glyphosate can also be applied to fields containing sensitive crops through the use of special applicators when the weeds extend above the crop canopy.

Pesticides are used and applied in a number of different ways. Some are applied to the soil well before the crop is planted, either as fumigants or as herbicides. Others, particularly insecticides, are applied as aerial sprays well after the crop has emerged. These differing application methods can affect the potential for drinking water contamination. Generally speaking, the potential for drinking water contamination is higher when more of the applied pesticide reaches the soil and when the pesticide is used at a time when there is either no crop or the crop is very young. A large, actively growing crop is able to reduce the potential for off-site movement of the pesticide through a number of mechanisms. Primary among these is the capability of a large crop to consume virtually all of the available water in the soil through transpiration. This lessens or eliminates the quantity of water that would be available to leach or runoff, thereby carrying the pesticide with it. Such water use can result in considerable direct uptake of the pesticide into the crop, helping to minimize off-site mobility. The crop also deflects and intercepts much of the incoming precipitation that might otherwise pond at the soil surface and subsequently run off the edge of the field. In addition, the presence of the crop often provides for a perfect soil environment in which a microbiologically diverse root zone can develop, enhancing the soil's degradative capacity.

The terminology generally used to differentiate whether a pesticide is applied before or after the crop has emerged from the ground is preemergent vs. postemergent. Postemergent pesticides are generally intended for plant foliage, either of the weeds or the crop. Preemergent pesticides become active as a result of their presence in the soil. Preemergent materials may be further subdivided as to whether they are left on the soil surface or are incorporated with a farming tool into the upper few inches of the soil profile. Certain pesticides are effective only when incorporated, either because they do not possess sufficient mobility to move down into the soil where they are needed or because they dissipate too rapidly due to volatility or photolysis when left exposed at the surface. Incorporation of moderately mobile pesticides into the soil can reduce the potential for runoff and therefore the potential for surface water contamination by the pesticide, but it often increases the potential for leaching and therefore groundwater contamination by the compound.

The hysteria that has sometimes accompanied the issue of pesticide occurrence in drinking water is not too difficult to rationalize. The public generally perceives that every chemical used by farmers must be acutely toxic, and everyone drinks at least some water. Of the liquids on Earth, the psychological bond to drinking water is rivaled only by mother's milk in terms of our reliance upon it as a continuous and untainted source. However, the high-pitched tenor of certain activists concerning pesticides in drinking water is clearly unwarranted. A good deal of public apprehension is caused by the misperception that the world is in the midst of a cancer epidemic directly linked to the many "poisons" we have introduced to the environment. The epidemiological facts are at odds with this perception. According to the American Medical Association (JAMA 1981), the only absolute age-adjusted increase in cancer incidence in the United States during the past 40 years has been an increase in lung cancer known to be due largely to cigarette smoking. While nearly 1 in 3 Americans is likely to develop some type of cancer, nearly all of these are a direct result of personal choices involving either smoking or a fat-laden diet.

Only 1 to 5 percent of all cancers can be attributed to environmental causes (Dahl and Peyto 1989).

Not all fresh water is of drinking quality and therefore potable. Small streams, ponds, and shallow groundwater are prevalent in the agricultural regions of the world, but none of these is routinely relied upon as a source of drinking water. Surface water sources of drinking water are nearly always publicly owned systems utilizing large rivers, reservoirs, or lakes. Extensive treatment of such surface water is generally required before it can be made suitable for human consumption. Water plants practice a number of techniques for removing sediment, adjusting pH, and adding such materials as chlorine and fluoride. As will be seen in Chapter 2, these techniques are generally not effective at removing the pesticides that can be present in the surface waters of areas draining agricultural regions.

Unlike surface water, deep groundwater is generally potable without any treatment. Public systems usually rely upon such deep (> 200 feet) wells, certainly below any depths where surface contaminants would be expected to be present at any significant concentrations. However, privately owned wells can be much shallower, and it is here that some of the distinctions between what is drinking water and what is not become somewhat blurred. In most areas of the world, including much of the United States, there are no permitting regulations concerning the construction of drinking water wells below a certain well diameter (for example, 2 inches). Thus, there is nothing to prevent (except perhaps a modicum of common sense) a landowner from pounding a well by hand into a 10-foot-deep surficial aquifer and using this for drinking water. Similarly, there are generally no regulations for the sealing and casing of deeper private wells to ensure that contaminated surface water cannot simply run down the sides of the well bore hole and thereby render it non-potable.

Given the rather obvious vulnerability of both shallow groundwater and surface water to contamination from land activities, the most surprising fact about the occurrence of pesticides in drinking water is that essentially nothing was known about it until 1979. In retrospect, this was probably due to two factors: (1) the unavailability until that time of reliable and sensitive analytical chemistry methods for detecting such materials in drinking water, and (2) an undue faith in the ability of soil to degrade or otherwise absorb the relatively minute initial concentrations of pesticides at the soil surface. As shown in the examples that follow, this faith (some might call it Orwellian blissful ignorance) was instantly and irreversibly broken in 1979 by notorious incidents, first involving DBCP in California and then aldicarb on Long Island, New York. The DBCP story, which just preceded the aldicarb discovery, will be examined first.

DBCP

DBCP (see Figure 1-1 for structure and properties) was developed in the 1950s under the trade name Nemagon to control certain nematodes and act as a soil fumigant. In 1977, chemical industry employees involved in the handling of the compound were found to have suffered significant impairment of their reproductive capabilities, prompting the California Department of Food and Agriculture to suspend the use of DBCP on selected agricultural crops (Wagenet, Hutson, and Biggar 1989). Cancellation of the use of the chemical nationwide by the U.S. Environmental Protection Agency (EPA, the Agency) was prompted due to the groundwater detections discovered in 1979.

In the laboratory, DBCP does not appear to be a chemical likely to leach into groundwater. As is obvious by its strong chemical odor, DBCP is quite volatile. This property of the molecule led most scientists and users to believe that when it was applied to soil it would last for a short time and then simply evaporate, leaving no significant residues for any subsequent off-site movement and contamination of drinking water.

This thinking was mostly true in the sense that at least 90 percent of the applied compound did evaporate under normal conditions. Nevertheless, a significant fraction of the applied compound remained in the soil profile after application. Though DBCP does not interact strongly with soil due to binding

DBCP

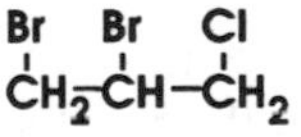

$$\underset{\substack{|\\CH_2}}{Br}\ \underset{\substack{|\\CH}}{Br}\ \underset{\substack{|\\CH_2}}{Cl}$$

$CH_2\text{-}CH\text{-}CH_2$

1,2 dibromo-3-chloropropane

Molecular Weight 236.36 g/mol

Half-life in soil, DT50: *200 days*

Soil Binding Coefficient, Koc: *65 L/kg*

Groundwater Ubiquity Score: *5.00*

Maximum Contaminant Level: *0.2 μg/L*

Water Solubility: *1000 mg/L*

Vapor Pressure: *0.8 mm Hg*

FIGURE 1-1.   Structure and properties of DBCP.

with organic matter or clay, it does possess a relatively high water solubility (1,200 mg/L). Thus, any water present in the soil dissolves a significant fraction of the compound.

In fact, according to Henry's Law from the science of thermodynamics, the presence of the water lowers the effective vapor pressure that the molecule can achieve. For an ideal solute, the reduction of vapor pressure and therefore volatility is in direct proportion to the ratio of the concentration in water to the water solubility of the compound. In the case of DBCP, which would be present at rates of approximately 1 LB/A after initial volatile losses had subsided, the concentration in the top 6 inches of soil is equivalent to 7.5 mg/L at a typical soil moisture content of 10 percent (v/v). Since the water solubility is 1,200 mg/L, this suggests a 160-fold reduction in the pure solute equilibrium vapor pressure of 0.8 mm Hg. What this means is that the compound's rate of dissipation would be influenced greatly by the presence of soil water, and any practices such as extensively watering the chemical into the soil profile would increase its persistence in at least two ways: (1) by lowering the effective vapor pressure and (2) by leaching chemical away from the soil surface.

Once washed into the soil profile, DBCP is quite persistent, because, unlike most pesticides now in use, it is not particularly biodegradable or susceptible to chemical hydrolysis. This combination of resistance to degradation and lack of interaction or sorption to soil is the key juxtaposition of properties leading to the potential for drinking water contamination.

Beginning in the mid-1960s, the use of DBCP became widespread throughout the San Joaquin river valley of California (see Figure 1-2). By far the most diverse and economically important agricultural region in the United States, this portion of California supports numerous crops, and although it does not suffer from some of the insect pests that frequent the Midwest, it does have more than its share of nematode problems. With such widespread use of the compound to combat these problems, it should not have been too surprising when the analysis of several well water samples in 1979 showed the occurrence of DBCP routinely at levels well above those thought to pose a risk to human health. DBCP has a maximum contaminant level established by the EPA of 0.2 μg/L.

Subsequent sampling throughout the remainder of the country never revealed the same levels of contamination that had been observed in California and hydrogeologically similar regions of Arizona. The precise reasons for this are not known, but it was probably due to the unique nature of the long history of intensive use of the compound in the Southwest, combined with the extensive watering that accompanied its applications. No other areas of the United States have the usually bountiful,

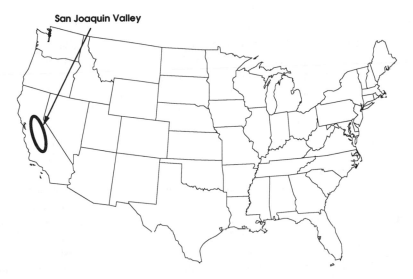

San Joaquin Valley

FIGURE 1-2.   Map showing areas reporting DBCP in drinking water.

cheap supplies of irrigation water that are available to farmers in these areas.

Despite a fairly limited data base of sampling, the state of California was able to determine that approximately 2,000 of the 8,000 wells sampled for the presence of DBCP contained the nematocide. The area of contaminated groundwater is thought to encompass 7,000 square miles of the San Joaquin Valley. It was used mainly in areas with sandy soils, including the east side of the San Joaquin Valley, but the actual acreage to which it was applied is not known because of the lack of accurate records. Typical annual application rates were extremely high by today's standards, ranging from 20 to 80 LB/A, with the total usage estimated to have been at around 3 million pounds annually at the time of cancellation in California in 1977.

Most of the groundwater samples containing DBCP had concentrations well below 100 μg/L. A good correlation was identified between the presence of high nitrate levels and the occurrence of DBCP in the wells. This was taken as an indication that the DBCP had been transported to the groundwater via percolating irrigation water. This hypothesis is consistent with the predictions of DBCP concentration profiles in soil using a computerized pesticide leaching model, LEACHM (Wagenet, Hutson, and Biggar 1989).

The incident with DBCP also led the EPA to investigate the possibility that other nematicides such as aldicarb and EDB might have found their way into groundwater. As will be seen in the case studies that follow, they did.

ALDICARB

Aldicarb (see Figure 1-3 for structure and properties), the active ingredient in Temik® brand Aldicarb Pesticide, has been used since the mid-1970s to control nematodes and insect pests in a number of important crops, including potatoes, oranges, bananas, peanuts, and cotton. It was originally developed by Union Carbide, but is now manufactured and marketed by Rhône-Poulenc, the French chemical giant who purchased several Union Carbide assets in 1987. This pesticide has been the subject of the longest EPA Special Review ever undertaken by the Agency. The review began with drinking (well) water contamination issues and has now expanded to include food residue concerns. Throughout this Special Review, the Agency has been breaking new ground on the manner in which pesticides are regulated in order to control the potential for drinking water contamination. As will be discussed at length in Chapter 5, the process has concluded with the EPA developing its Pesticides in Ground Water Strategy, based almost entirely on the model for how it plans to deal with aldicarb. Thus, it is worth gaining some perspective on the history of aldicarb in groundwater.

Applied as granules and then incorporated into the soil profile, moisture releases aldicarb from the granule and a large number of chemical and biological processes are set in motion that eventually result in the loss of

ALDICARB

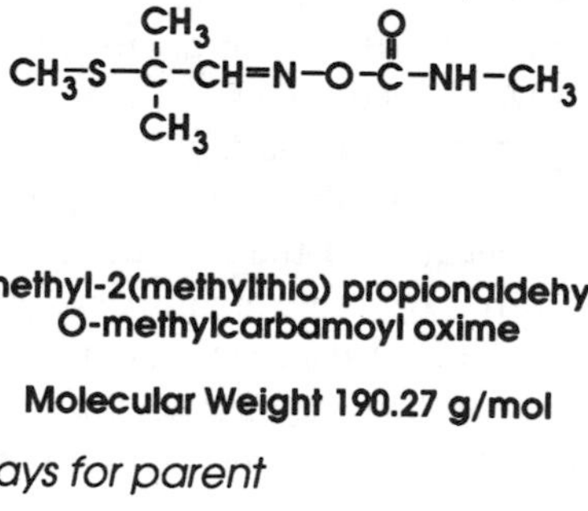

$$CH_3-S-\underset{\underset{CH_3}{|}}{\overset{\overset{CH_3}{|}}{C}}-CH=N-O-\overset{\overset{O}{\|}}{C}-NH-CH_3$$

2-methyl-2(methylthio) propionaldehyde
O-methylcarbamoyl oxime

Molecular Weight 190.27 g/mol

Half-life in soil, DT50: *26.5 days for parent*

Soil Binding Coefficient, Koc: *23 L/kg*

Groundwater Ubiquity Score: *3.75*

Maximum Contaminant Level: *10 μg/L*

Water Solubility: *6000 mg/L*

Vapor Pressure: *0.00008 mm Hg*

FIGURE 1-3.   Structure and properties of aldicarb.

biological activity. As carbamates, aldicarb and two of its soil degradates, aldicarb sulfoxide and aldicarb sulfone, are toxic to many important crop pests and to mammals, including humans. However, the dilution of the compound in soil and water leads to the occurrence of aldicarb residues in water and treated crops that are in general not biologically relevant to humans consuming the affected materials.

One important exception to this allegedly occurred in 1985 in California, in which watermelons were treated by Temik illegally, leading to considerably higher residues than normally seen in food crops and a number of sicknesses among particularly sensitive persons consuming the tainted fruit. Watermelons, by virtue of their rapid consumption of soil water, effectively concentrated residues of the aldicarb in their fleshy tissue. This incident does point out the propensity by aldicarb and its carbamate degradates to move freely with soil water in whatever direction the laws of physics tell the water to go.

New York's Long Island Incident

During the 1970s, much of New York's Long Island (see Figure 1-4) was used to produce potatoes. The sandy soils and plentiful rains that produced the bountiful crops of potatoes also led to a potential for vertical movement of water through the soil profile. The use of aldicarb on the potatoes had

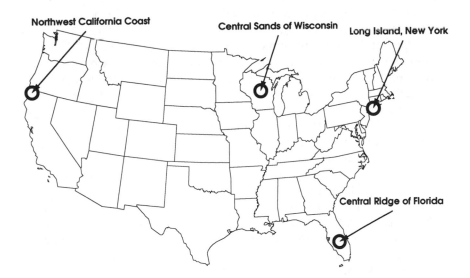

FIGURE 1-4.   Map showing areas reporting aldicarb in drinking water.

resulted in huge gains in yield, due primarily to more consistent control of the most devastating pest in that region, the Colorado potato beetle.

Following the examination of results of a water pollution study assessing the possible impact of pesticides in Suffolk County, the Cooperative Extension Service there recommended, in 1976, that the groundwater be sampled and analyzed for the occurrence of carbamate pesticides such as aldicarb. For lack of funding, this recommendation was not acted upon until August 1979, when samples from shallow wells located at Cornell University's Long Island Horticultural Research Laboratory at Riverhead indicated that the carbamate degradates of aldicarb (aldicarb sulfoxide and aldicarb sulfone) had moved through the root zone and into groundwater.

This finding spawned a small monitoring study between August 1979 and March 1980 during which approximately 270 wells were sampled on eastern Long Island. Aldicarb concentrations in 61 of these wells exceeded the health-based guideline of 7 μg/L that had been established by the state of New York for total residues of aldicarb and its carbamate degradates in drinking water. One fact about this drinking water standard worth noting is that it treats all three aldicarb-derived carbamates as equivalent, additive species despite the fact that aldicarb sulfone is approximately 25 times less toxic than either of the other two compounds. It is also worth noting that parent aldicarb is practically never found in groundwater, except in the time period immediately after application. It was these widely reported findings of aldicarb residues above the health-based guidelines in about a quarter of the sampled wells so close to the media moguls of New York City that helped sensationalize the story of drinking water contamination by pesticides.

Typical applications of aldicarb on Long Island were 5 LB/A of active ingredient at the spring planting followed by an additional 1 to 2 LB/A as a side-dressing at mid-season. Applications such as these were made on virtually all of the approximately 22,000 acres of potato fields in Suffolk County between 1975 and 1979. These use rates were significantly higher than the national use rates at the time of approximately 3 LB/A at planting. The higher rate of 5 LB/A at planting was required in order to obtain effective golden nematode control and the side-dressing was subsequently added in order to obtain useful control of the Colorado potato beetle.

In response to the initial findings of the study concluded in March 1980, an extensive well sampling program was initiated to evaluate the impact of aldicarb use on the water quality in all potable wells near potato farming operations. An intensive monitoring program was conducted from April to June 1980 that included the collection of water samples from approximately 8,000 wells, most of which were privately owned. Sampling was performed by the Suffolk County Department of Health Services, and chemical analysis of the water samples was performed by Union Carbide. Just over 1,000 of the wells in this first major survey had aldicarb residues in excess of the 7 μg/L standard. Another thousand had detectable residues ($>$ 1 μg/L) but were below the guideline. Extensive analysis of these data demonstrated that the highest concentrations were found next to the previously treated fields and that a distance of 2,500 feet was sufficient for residues to decline to undetectable levels.

Individual well owners within Suffolk County were informed of the results of the testing and were cautioned not to use the water for cooking or drinking purposes whenever residues exceeded the state guideline of 7 μg/L. Union Carbide offered, at the company's expense, granular activated carbon (GAC) filters to any well owner with such residues. The units continue to be recharged with fresh GAC beds as needed every 9 to 15 months, again at company expense. Monitoring continues in these areas and as residues fall below the guideline the GAC units are eventually removed.

Besides domestic wells, several larger municipal wells in Suffolk County were found to have aldicarb residues above the 7 μg/L state guideline. Considerably larger GAC treatment systems were required for these wells and were more expensive than the several hundreds of dollars expended on each of the household units.

The main factors leading to the pervasive nature of aldicarb contamination of groundwater in Suffolk County were the high rates of application, the heavy spring rains, the very permeable soils, the acid soil conditions, and the occurrence of so many shallow wells near the treatment areas. The pH of Suffolk County groundwater is between 5 and 6, and numerous research studies have demonstrated that aldicarb and its carbamate degradates exhibit a relative maximum in persistence at exactly this range. This combination of unfortunate circumstances could not have been completely predicted beforehand, but the Suffolk County experience is one that will hopefully teach some useful lessons in preventing the recurrence of such an incident.

There were various attempts to develop new ways of using aldicarb on Long Island at different times, at lower rates, or in controlled release formulations, but none of these proved to be effective, and it was finally concluded that it would be virtually impossible to use aldicarb on Long Island in such a way so as to avoid causing groundwater residues above the state's 7 μg/L guideline.

Some of the research conducted on Long Island included attempts to predict where the aldicarb residues will eventually reside and where they will eventually go. In 1982, it was predicted by the Suffolk County Health Department that it will take approximately 100 years for the East End's groundwater system to purge itself of the pesticide. This prediction was based on the extremely conservative assumption that the residues would not degrade at all and would simply decrease in concentration due to advection and dispersion. However, even when an observed half-life of 3 years is incorporated into these models, it would seem that several decades will be necessary for all wells to be freed of residues in excess of the guideline.

A complex three-dimensional particle tracking model of the recharge on Long Island, New York has been developed (Buxton et al. 1991). It was used to project future recharge areas on the island given current prediction about future population growth and water use, and could be used to generate more precise predictions about the future course of contamination there.

Wisconsin's Central Sands

In 1981, Union Carbide and various Wisconsin state agencies began an extensive groundwater sampling program in the Central Sands region of Wisconsin (see Figure 1-4), which has properties in many ways similar to the conditions prevailing on Long Island: extensive aldicarb use and relatively high susceptibility to leaching. Sites for sampling were selected based on areas of known nitrate contamination, 2 years of consecutive aldicarb use, sandy and acidic soils with low organic matter content, shallow water table, and the use of irrigation. Application of these criteria resulted in the selection of approximately 360 samples, of which 17 had aldicarb residues above the 10 μ1g/L Wisconsin state guideline.

This discovery was a major blow to Wisconsin potato growers, who considered aldicarb an extremely valuable pesticide. Use of the material had resulted in higher yields of 15 to 20 percent due to improved control of nematodes, Colorado potato beetles, aphids, and leafhoppers. Severe restrictions on aldicarb use were put in place as a result of the 1981 findings. The pesticide became a Restricted Use material, meaning that only state-certified pesticide applicators or persons under their direct supervision could apply the material. The timing of applications was changed such that they would be made from 4 to 6 weeks after planting, rather than at the time of planting. Applications were further restricted such that the same field could not be retreated the year after a treatment had been made, and application rates were reduced to a maximum of 2 LB/A, rather than the 3 LB/A maximum rate that had been on the label. This rate reduction forced Union Carbide to remove nematode control claims from the labeling of the product. Special regulations were put in place to control the use of aldicarb in Wisconsin. These are discussed in Chapter 5.

In laboratory studies of the mechanisms for the degradation of aldicarb and its carbamate degradates by researchers at the University of Wisconsin (Lightfoot et al. 1987), it was found that soil enhances the hydrolytic degradation, probably through simple surface catalysis. Temperature and pH were also identified as important factors in both soil and water, and no reduction of aldicarb sulfoxide or aldicarb sulfone back to the less oxidized forms could be induced.

In a study of aldicarb degradation rates at three depths in a Wisconsin Central Sand Plain aquifer (Kraft and Helmke 1991), it was found that the degradation rates were similar for all three depths of 0.5, 5.0, and 10.0 m below the water table. The results are consistent with the hypothesis that aldicarb degradation within the saturated zone is primarily a chemical process and does not depend to any significant extent on microbial processes, which are much more rapid in the shallower depths below the water table surface.

Florida

In 1982, the aldicarb story shifted to Florida, where roughly 10 percent of the 850,000 acres of citrus were being treated annually with the compound to control citrus nematodes and other pests. At that time up to 10 LB/A were being applied annually on each field, usually as two applications. In August 1982, residues of aldicarb were found in wells near an experimental agricultural site, and follow-up sampling of other wells near commercial production areas showed several detections. The state acted quickly to establish a $10\,\mu g/L$ health advisory level and impose a 1-year ban on most uses of the compound. Research was also initiated between state agricultural experts and Union Carbide to develop methods for using the compound without negatively impacting drinking water quality. The import of this research was a set of new application rules to come into effect after the 1983 ban. Maximum application rates were limited to 5 LB/A, and the use on citrus was limited to a time window between January 1 and April 30, which limited the exposure of freshly applied compound to the very heavy late spring and summer rains. In addition, there was the first invocation of a well set-back rule, in which no use of aldicarb was permitted within 300 feet of any drinking water well throughout the state. On the central-ridge portion of the state (see Figure 1-4), where hills and other topographic features lead to much faster lateral groundwater velocities, the 300-foot set-back was subsequently increased to 600 and then 1,000 feet, based on field research and computer modeling studies (Jones et al. 1987).

The results of several field monitoring studies conducted in Florida demonstrated that application to potatoes and bedded citrus groves in coastal areas resulted in some residues reaching shallow, surficial aquifers, but degradation then occurred rather rapidly. However, in the central sandy ridge region, the higher recharge rates, slightly acidic sandy soils, and low organic matter all contributed to the greater persistence of aldicarb residues in the groundwater. These combined with the aforementioned faster lateral groundwater velocities to make it necessary to increase the 300-foot set-back to the 1,000-foot distance. Modeling work by the EPA (Dean and Atwood 1987) has confirmed the utility of this set-back distance.

Other States

Follow-up sampling throughout the rest of the aldicarb use area during the 1980s has shown sporadic detections in many states, most often in areas that have soils and climatic conditions similar to Long Island—sands and heavy rains or irrigation. These have been primarily limited to other areas of potato production in the Northeast. In addition, a few contaminated wells have been identified in the Far West along the Pacific coast (associated with lily bulb production). The observation that significant movement of aldicarb occurs primarily in sandy soils was corroborated by a study of aldicarb movement in different soil types (Awad, Kilgore, and Winterlin 1984), in which it was found that the highest levels of aldicarb residues in soil column leachate came from sandy, sandy-loam, and peat soils, whereas the lowest amounts were found in clay and loam soils.

As a result of aldicarb's demonstrated potential to enter groundwater in areas of sandy soils, the registrant, now Rhône-Poulenc Ag Company, has developed an extensive list of soil names on the label where the use of the compound is restricted. The restrictions specify "well set-backs" of either 300, 500, or 1,000 feet that are protective of shallow drinking water wells in cases where leaching into groundwater is likely to occur based on the susceptibility of local soil conditions. Specific use windows similar to the ones established in Florida on citrus were developed for other parts of the country. Extensive field research and computer modeling efforts on the part of Russell Jones of Union Carbide and then Rhône-Poulenc was behind the establishment of most of these recommendations (Jones, Black, and Estes 1986; Jones 1987; Jones, Hornsby, and Rao 1988; Jones and Bostian 1989). The validity of these well set-back distances under French conditions was explored in one study in which the conventional equilibrium theory of the conventional gas chromatography model was applied to the movement of aldicarb residues through aquifer materials (Mestres et al. 1987). Despite the obvious over-simplifications of this approach, the results obtained were similar in magnitude to what has been reported using more sophisticated approaches in the United States. A well set-back distance of 50 meters was said to be adequate for a moderately rapid horizontal water velocity of 0.6 m/day.

Now, as a final step in the regulatory process, the EPA has considered mandating that any state wishing to permit the use of aldicarb shall develop and use state management plans (SMPs). These SMPs specify the regulations on aldicarb use that will be necessary in the individual states to prevent contamination under their local conditions. In adopting this regulatory strategy, the Agency acknowledges that each state's situation is unique regarding the manner in which the compound is used and the vulnerability of the local drinking water supplies. In the EPA's national evaluation of aldicarb leaching potential in the 11 groundwater regions defined by Heath (Lorber et al. 1989), the use of aldicarb on potatoes in the north central and northeastern parts of the country was given the highest concern rating. High leaching potential ratings were also given to uses on citrus in Florida.

Besides leading to the development of pioneering regulatory strategies, the various incidents of groundwater contamination by aldicarb have inspired a number of researchers to study various aspects of aldicarb's environmental fate processes in more depth. Some of the findings that have broadly applicable significance are included here.

The richest area of research into the question of aldicarb movement has been the development of computer models such as PRZM (see Chapter 4) to describe the process. Researchers have found that the predictions of pesticide transport models are highly sensitive to the values used for degradation rate constants and sorption coefficients. This was illustrated dramatically by Villeneuve et al. (1988) in the case of predicting aldicarb leachate with PRZM, for which a 15-percent variation in the degradation rate constant or a 24-percent variation in the sorption coefficient each led to 100-percent uncertainty in the cumulative quantity of aldicarb leached below the root zone or the dissolved concentration of aldicarb in the leachate.

In a study on the field validation of a groundwater model using aldicarb movement in Wisconsin, it was concluded that rigorous field validation and calibration were often impossible due to uncertainties in input parameters (Anderson 1986). Nevertheless, it was emphasized that an attempt should always be made to demonstrate that a model is capable of predicting concentrations measured in the field. Excellent agreement between PRZM model predictions of the leading edge of the leaching front for aldicarb have been reported (Jones, Black, and Estes 1986), although concentration profiles were not well represented. Relatively good agreement between RUSTIC model predictions and observed aldicarb behavior was obtained for the Wickham site on Long Island, New York (Huyakorn, Kool, and Wadsworth 1988). The RUSTIC model, developed for but now abandoned by the EPA, tied PRZM model predictions to a deeper unsaturated zone model and a two-dimensional saturated zone model (see Chapter 4).

The question of environmental effects on aldicarb fate and movement has also been the subject of intensive research. The pH and temperature dependence of aldicarb movement through laboratory soil columns was recently described (Lemley, Wagenet, and Zhong 1988). These researchers found that sorption to soil decreased at higher temperature, but not enough to overcome the beneficial effects of higher temperatures on the hydrolysis rates of aldicarb and its carbamate degradates in soil. The activation energy reported for the hydrolytic degradation of aldicarb sulfone was 15 kcal/mol, almost identical to the value found in aqueous solution. Formation of the carbamate oxidation products of aldicarb are not observed under anaerobic conditions (Ou, Edvardsson, and Rao 1985). Degradation of aldicarb and its carbamate degradates appears to slow down slightly in deeper soils typical of Florida (Ou et al. 1985; Ou et al. 1988).

EDB

Following the DBCP incidents and the discovery of aldicarb in a few of Florida's wells, another nematocide was investigated in the state: ethylene dibromide or EDB (see Figure 1-5 for structure and properties). EDB is structurally very similar to DBCP, and has very similar properties and use patterns. EDB was used to control burrowing nematodes on citrus in Florida. It was applied for over 15 years by growers and by the state to create barrier zones, stopping the spread of nematodes. In addition, the pesticide was used extensively on golf courses and on such crops as peanuts and soybeans. The cancellation of most uses of DBCP in the late 1970s had increased the use of

EDB

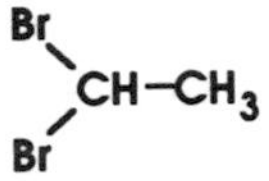

$$Br\diagdown\atop Br\diagup CH-CH_3$$

1-2-dibromoethane

Molecular Weight 187.88 g/mol

Half-life in soil, DT50: *1500 days*

Soil Binding Coefficient, Koc: *80 L/kg*

Groundwater Ubiquity Score: *6.60*

Maximum Contaminant Level: *0.05 µg/L*

Water Solubility: *3400 mg/L*

Vapor Pressure: *12 mm Hg*

FIGURE 1-5.   Structure and properties of EDB.

EDB. EDB had labeled uses at application rates as high as 50 LB/A and was sometimes injected to 4-foot depths in the soil profile.

Just as with DBCP, the high volatility of the compound led many to believe that it could not possibly contaminate groundwater. Unfortunately, as pointed out in the case of DBCP, the relatively high water solubility of the compound means that the effective vapor pressure in wet soils is dramatically reduced as a direct result of Henry's Law. It is also apparent that EDB does not biodegrade at any significant rate, and the half-life in groundwater is now thought to be on the order of many years. Entrapment of EDB within the intra-particle micropores of soil has been cited as one explanation for the observed persistence of the compound (Steinberg, Pignatello, Sawhney 1987).

The Florida testing program for EDB began in earnest in August 1983. Almost immediately, concentrations many times above the state-established health guideline of 0.1 μg/L were observed, and the state banned its further use just one month later (September 1983). As of late 1984, 36 municipal wells were known to have residues above the state guideline, and over 600 private (generally much shallower) wells were similarly contaminated. Activated carbon filters were installed by the state on all contaminated wells, except when it was feasible to connect the well owner to an alternative supply.

Besides being found in Florida, EDB has been reported in wells in Georgia, California, South Carolina (see Figure 1-6), and Hawaii, although it is now questionable whether the detection in Hawaii was a result of normal

FIGURE 1-6.   Map showing areas reporting EDB in drinking water.

agricultural use. Soil coring data in California showed that residues at a concentration of 1 μg/L had leached down to 10 meters beneath the soil surface. According to the EPA (Cohen et al. 1984), it was this soil coring result that was the driving force behind the emergency suspension of all soil fumigant uses of EDB in 1983. A thorough review of the entire regulatory history of EDB is available from Pignatello and Cohen (1990).

ATRAZINE

After the detection of aldicarb, DBCP, and EDB in well water, attention shifted to other possible contaminants. Atrazine (see Figure 1-7 for structure and properties) was a natural choice because, along with alachlor (see next section), it was among the top two active ingredients in terms of the number of pounds applied to the soil each year in the United States. It had been used for many years and was known to be fairly persistent, to the extent that it is known to cause "carry-over" problems in certain areas of the country. Carry-over refers to the ability of any pesticide, but generally only herbicides, to exhibit continuing activity in a crop planted after the originally treated

ATRAZINE

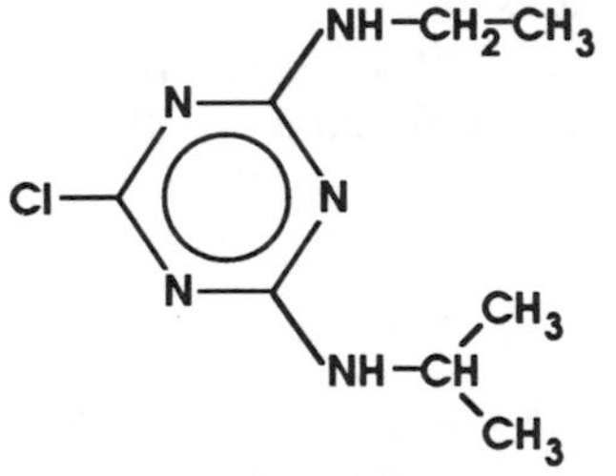

2-chloro-4-(ethylamino)-6-isopropylamino
-s-triazine

Molecular Weight 215.69 g/mol

Half-life in soil, DT50: *80 days*

Soil Binding Coefficient, Koc: *160 L/kg*

Groundwater Ubiquity Score: *3.44*

Maximum Contaminant Level: *3 μg/L*

Water Solubility: *33 mg/L*

Vapor Pressure: *0.0000003 mm Hg*

FIGURE 1-7.   Structure and properties of atrazine.

crop has been harvested. Both purely chemical and microbially mediated processes have been identified in reactions involving atrazine metabolism in soil.

Atrazine is a triazine herbicide that has been widely used on sorghum and field corn to control grass and broad-leafed weeds for several decades. Because of its low water solubility (33 mg/L) relative to the three pesticides just mentioned, many researchers had discounted the possibility that atrazine could leach and thereby enter groundwater. This underlying assumption of equating low water solubility with low mobility in soil is common among most biologists and chemists working in the area of agricultural chemicals. However, current data suggest this assumption is very misleading. As discussed more thoroughly in Chapter 4, it is now thought that only those compounds having water solubility below about 1 mg/L are really insoluble enough not to pose a possible water contamination threat.

Monitoring

Atrazine is very widely used throughout the United States and especially in the midwestern states where corn is the dominant crop. The EPA estimated in 1988 that atrazine use nationwide was approximately 88 million pounds annually from 1985 to 1987 (Barrett and Williams 1989). As early as the mid-1970s, the USGS had already measured atrazine in many of the nation's rivers (Gilliom, Alexander, and Smith 1985). Compilations of available groundwater data through 1988 by the EPA indicated that approximately 4 percent of the 15,000 wells sampled through that year contained atrazine. Since detection limits vary and the sampling protocols were generally not designed to provide nationwide estimates, it is unclear what this figure really means; however, it is clear that a rather large number of wells do contain traces of atrazine. Fortunately, very few (perhaps only a fraction of a percent of the wells sampled) contained residues above the MCL currently established by the EPA for the herbicide: 3 μg/L.

In Nebraska, an extensive survey for the occurrence of atrazine in private wells has been underway now for several years (Jacobs 1988). The goal of the survey is to collect samples from about 2,000 domestic wells during a 5-year period. Thus far, atrazine has been the most commonly found analyte, occurring in about 6 percent of the samples at levels above the state's analytical reporting limit of 0.05 μg/L. The analytical methodology employed by the state includes the confirmation of all positive detections through a second, independent analytical method. Most of the detections were in the Platte River Valley region of south central Nebraska and were located in areas of the most intensive corn and sorghum production. Other factors leading to the occurrence of atrazine in the groundwaters of this region were the highly permeable and excessively well-drained soils and subsoils, intensive irrigation, and heavy reliance upon atrazine for weed control.

Sampling in Kansas for the occurrence of pesticides in groundwater has consistently revealed atrazine to be the most common contaminant, but generally in somewhat fewer than 4 percent of the samples collected (Robbins 1988). This relatively small number of wells was surprising because sampling had been confined only to quite vulnerable regions.

In Iowa, atrazine was detected in 18 percent of a set of relatively vulnerable municipal wells located mainly in the western third of the state (Detroy, Hunt, and Holeb 1988). Of the pesticides examined, atrazine was by far the most commonly observed analyte. The next closest was metolachlor, which was detected in 5 percent of the wells. There was a fairly strong association between the occurrence of atrazine and well depth in this study. Atrazine occurred in 21 percent of the wells less than 100 feet deep and in only 10 percent of the wells deeper than 100 feet. The median well depth for these municipal systems was only 65 feet, suggesting one reason for the relatively high proportion of wells identified in this survey.

In a complete survey of all 800 municipal systems throughout the state of Iowa, including both surface and groundwater systems, approximately 10 percent contained detectable residues of atrazine (Iowa DNR 1988). The authors of this study felt that nearby pesticide storage facilities may have been the source of the compound in many of these cases. Cyanazine, another triazine herbicide, was the next most commonly observed pesticide in the study, occurring in about 5 percent of the samples. As in the more restrictive study of municipal systems by Detroy, Hunt, and Holeb, there was a strong association between well depth and the occurrence of pesticides, with 16 percent of the wells less than 100 feet deep having at least one pesticide, and only 5 percent of the wells deeper than 100 feet deep having detectable levels.

In the karst areas of northeast Iowa, atrazine has been even more commonly found (see Figure 1-8). The Big Spring Basin in Clayton County is an area that has been the subject of intensive scrutiny over the past several years, mainly under the research direction of George Hallberg. The major crop in the basin is corn production, which accounts for over half the land area. The county is riddled with sink holes and other features common to karst topography. This leads to the direct contamination of groundwater by surface runoff. Most of the shallow monitoring wells show detectable levels of atrazine, although concentrations in excess of the current MCL (3 μg/L) are extremely rare and short-lived. Despite the preponderance of low levels of atrazine in surface and groundwaters of this part of the state, there is no evidence that algal populations have shown a significant response (Hersh and Crumpton 1989). Populations of algae collected from Big Spring, which

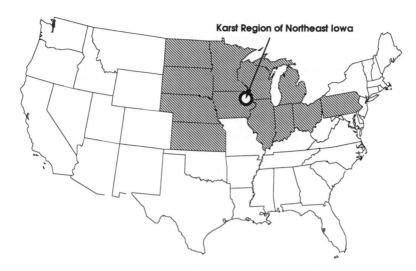

FIGURE 1-8.   Map showing areas reporting atrazine in drinking water.

routinely contains concentrations of atrazine of a few tenths of a microgram per liter, were not more resistant to the herbicide than algae derived from a more pristine source (Osage Spring). Both samples exhibited a wide range of susceptibility to the herbicide.

In Floyd and Mitchell Counties, somewhat farther to the west of Clayton County, the land is typified for the occurrence of incipient karst conditions, meaning that limestone is beneath the soil and is dissolving but not in areas large enough to cause exposed sink holes. Instead, channels lurk just beneath the soil surface and are capable of serving as conduits for the movement of pesticides into groundwater used as a source of drinking in these rural communities. A systematic scheme was used by two teams of researchers (Libra et al. 1984; Morgan et al. 1988) in an attempt to provide reliable estimates of the proportion of the entire population of wells in these two counties likely to contain atrazine and other pesticides. With a detection limit of 0.13 μg/L, atrazine was detected at least once in 43 percent of the drinking water wells during 1 year of quarterly sampling. The average percentage of wells having detectable levels of pesticides at any given time was 27 percent for atrazine, and less than 10 percent for the other detected pesticides (alachlor, metolachlor, cyanazine, metribuzin, and propachlor). The detection limits of the analytical methodology were considerably lower for these other materials—approximately 0.01 μg/L for each.

Significantly, atrazine was the most common drinking water contaminant in this survey despite the fact that other pesticides (alachlor, cyanazine, metolachlor) are just as widely used and the analytical method was considerably less sensitive for atrazine. Besides pointing out the relatively higher propensity for atrazine to be a contaminant of groundwater, these data also showed the importance of hydrogeologic setting in determining the likelihood of observing pesticides in well water. The wells in Floyd and Mitchell Counties were divided into four different classifications based on hydrogeology: (1) karst, characterized by the occurrence of visible sinkholes, (2) incipient karst, having very shallow bedrock with closed soil-filled depressions, (3) shallow bedrock, less than 50 feet deep, and (4) deep bedrock, more than 50 feet deep. The respective frequencies of pesticide detection for the four categories were 92, 70, 65, and 41 percent, indicating again how important the filtration of water through soil is in removing pesticides from water before potable supplies are reached.

In Illinois, the state Environmental Protection Agency has been analyzing public water supply wells for the occurrence of atrazine and other commonly used pesticides since 1985 (Good and Taylor 1987). The sampling network was established to assess the water quality in wells in the principal sand and gravel aquifers of the state. Less than 1 percent of the wells have been found to contain any pesticides, including atrazine. These results again show that deeper wells are much less likely to contain atrazine or other commonly used pesticides, even when located in regions of high and continuous use.

Most monitoring studies have been limited to atrazine and have not included its soil metabolites. The validity of this approach has been questioned severely by one laboratory study. In an experiment involving the formation and transport of deethylatrazine in the soil and vadose zone (Adams and Thurman 1991), the authors found that this soil metabolite of atrazine was the major degradation product of atrazine identified in soil water, and it appeared to enter the underlying aquifer at a concentration of 5 μg/L, which was greater than the concentration of atrazine entering the aquifer. The soil sorption coefficients of two other atrazine soil metabolites were found to be even higher than that of atrazine (Clay and Koskinen 1990), suggesting they would not be as mobile as either atrazine or deethylatrazine. The validity of extrapolating such laboratory data to field conditions has been supported in one study (Winkelmann and Klaine 1991a and 1991b), in which atrazine metabolism in a western Tennessee soil under laboratory conditions using minimally disturbed soil samples was said to be identical to what occurred in the field.

ALACHLOR

Alachlor (see Figure 1-9 for structure and properties) is an acetanilide herbicide that has been used to control mainly grassy weeds in corn, soybeans, peanuts, and a few other crops such as dry beans since 1969. It is the active

ALACHLOR

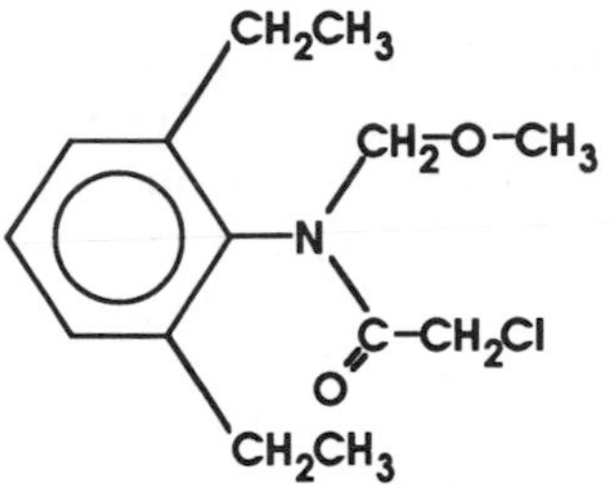

CH$_2$CH$_3$

CH$_2$-O-CH$_3$

N

C-CH$_2$Cl

O

CH$_2$CH$_3$

2-chloro-2',6'-diethyl-N-(methoxymethyl)
acetanilide

Molecular Weight 269.77 g/mol

Half-life in soil, DT50: *15 days*

Soil Binding Coefficient, Koc: *170 L/kg*

Groundwater Ubiquity Score: *2.11*

Maximum Contaminant Level: *2 µg/L*

Water Solubility: *240 mg/L*

Vapor Pressure: *0.000014 mm Hg*

FIGURE 1-9.   Structure and properties of alachlor.

ingredient in Lasso® and other herbicides by Monsanto. The finding in the early 1980s of tumors in small laboratory mammals that were fed large doses of alachlor prompted a careful examination of the potential for alachlor to occur in drinking water supplies derived from ground and surface sources. In some ways, alachlor was initially viewed by the EPA as having a greater potential for water contamination than atrazine because of its higher water solubility (240 mg/L), higher use rate (generally 2 times that of atrazine), and the larger number of crops on which the herbicide has utility. However, the critical difference between alachlor and atrazine with regard to their respective environmental fate profile was the much faster degradation rate of alachlor in soil. Alachlor, having a soil half-life on the order of 2 weeks, is much less persistent than atrazine.

In a study of dissipation rates of three acetanilide herbicides, including alachlor in soil (Beestman and Deming 1974), it was found each compound was approximately 50 times more stable in sterilized soil than in viable soil, suggesting immediately the importance of microbial activity in the dissipation of these compounds. Dissipation was found to observe linear, first-order dissipation kinetics, and there was an apparent association of water solubility with dissipation rate—the more soluble the herbicide, the faster it was degraded by microbes. In addition to microbial degradation, the herbicides were found to volatilize at appreciable rates when the soil was kept moist and a turbulent air surface (windy) was maintained.

Microbes and soil fungi were shown in another study to be involved in the dissipation of alachlor from surface soils (Tiedje and Hagedorn 1975). The aerobic and anaerobic degradation of alachlor in soil samples collected in southwest Georgia indicates that the rate of soil degradation of the molecule would slow down appreciably as it leaches down through the profile to an aquifer (Pothuluri et al. 1990).

As an apparent result of its rapid rate of degradation in surface soils, relatively few incidents of drinking water contamination by alachlor have been reported (see Figure 1-10). This is true despite the fact that in several individual years during the 1980s more pounds of alachlor were applied than of any other individual pesticide. It is still among the largest volume materials in the United States. The EPA recently established a health-based maximum contaminant level of 2 μg/L for alachlor, indicating that such levels of the compound in drinking water do not pose an unreasonable risk to human health. The episodes of monitoring for alachlor now discussed are exclusive of the much broader national surveys covered in Chapter 2.

In Nebraska, alachlor was found in 2 out of 14 irrigation wells sampled at low concentrations of less than 0.1 μg/L, despite the fact that Lasso® had been heavily used in the area and nitrates routinely exceeded 10 mg/L in the same wells (Jacobs 1988). In another Nebraska study, alachlor was found

FIGURE 1-10.   Map showing areas reporting alachlor in drinking water.

in only one of 35 wells sampled in a vulnerable alluvial region near the Platte River.

In surveys of northeastern Iowa surface and groundwater reported by Libra et al. (1984), alachlor was found in one well at concentrations near 20 μg/L, but in only a small percentage of the remaining samples and never at concentrations above 1 μg/L.

In Delaware, concentrations of alachlor in shallow monitoring wells directly beneath treated fields were reported as high as 15 μg/L, but no drinking water samples contained measurable traces of the compound (Ritter 1990).

A detailed study of computer model development and comparison with observed field behavior of alachlor in the Rathbun Reservoir of south central Iowa suggested that laboratory-derived estimates of the first-order dissipation rate constant were sufficient to predict observed field behavior (Schnoor et al. 1981). Predicted and observed concentrations in the reservoir reached a peak of 0.3 μg/L and declined below detectable levels within a few weeks.

In a study of alachlor behavior in North Carolina estuaries adjacent to treated farmland (Eisenriech and Sandstrom 1987), concentrations of dissolved and particulate alachlor were measured during and following spring application of the compound. Alachlor was not found in the estuary waters before application, but increased to 0.05 μg/L (all dissolved) after application but prior to runoff. Concentrations in the headwaters of the creek feeding the estuary reached 0.3 μg/L at this same time. Subsequent runoff events increased alachlor concentrations in the estuary to as high as 15 μg/L, but these levels were very short-lived and concentrations rapidly reached undetectable levels (< 0.004 μg/L) after a few weeks.

As with atrazine, most monitoring studies for alachlor have been limited to the parent compound. In a study of the fate and transport of alachlor, atrazine, and metolachlor in large soil columns utilizing radiolabeled materials in the laboratory (Alhajar, Simsiman, and Chesters 1990), it was found that the order of herbicide mobility was alachlor > metolachlor >> atrazine. Virtually all of the movement observed in these studies was due to the soil degradates of the three herbicides, and the results tend to reflect which of the three parent compounds was degraded the fastest to more mobile materials. As many as 8 to 12 alachlor degradates and 2 to 6 metolachlor degradates appeared in the leachate from the columns.

NITRATES

Although not a pesticide, nitrate ion is a common contaminant of drinking water closely associated with agricultural activities. The application of nitrogen as inorganic fertilizer to many crops is essential to the economic viability of agriculture throughout the world. Nitrates have a variety of uses other than as inorganic fertilizers. They are also used in the manufacture of explosives, in glassmaking, and as a heat-transfer fluid and a heat-storage medium for solar heating applications (EPA 1988).

Nitrate is a naturally occurring ion that makes up part of the nitrogen cycle. Wastes containing organic nitrogen enter the soil and are decomposed to ammonia, which is subsequently oxidized to nitrite ion and finally nitrate. Nitrate ion, like soil, carries a negative charge and is therefore very mobile in soil and essentially moves freely in whatever direction and velocity the soil water happens to be moving.

Nitrates occur naturally in a number of foods, especially vegetables. They are also added to meat products as preservatives. Levels of nitrate ion in surface and groundwater unaffected by human activity are generally less than 1 mg/L. Contamination from excessive nitrogenous fertilizer applications, feedlots, and septic systems have raised concentrations of nitrate in drinking water above the 10 mg/L MCL for nitrate in many areas of the country. The EPA recently estimated that 4.5 million people in the United States drink ground (well) water containing > 10 mg/L nitrate, a figure that dwarfs the number likely to be exposed to any pesticide above its MCL (EPA 1992).

The 10 mg/L MCL for nitrate ion is based on a toxicological pathway whereby nitrate is reduced in the body to nitrite ion, which subsequently reacts with hemoglobin to form methemoglobin. This altered form of the blood protein is unable to transport oxygen and can therefore lead to asphyxia in severe cases. No other significant toxic effects have been observed at relevant concentrations, and even at exposure levels of 111 mg/L there was a lack of any observed adverse affect on infants (Craun, Greathouse, and Gunderson 1981).

Research on nitrate movement through soil has generally shown that the amount of nitrate leaching is closely tied to the amounts applied in relation to precipitation and irrigation in the weeks just after application. Some of this research has been based on the use of specially designed lysimeters (Soileau and Hauck 1986). Crops, such as potatoes, that are grown in sandy soils are especially likely sources of leaching nitrate ion (Milburn et al. 1990). Besides leaching as a uniform front through soil, nitrate ion can move very rapidly through soil cracks and fissures. Such preferential flow patterns through macropores in a midwestern Nicollet soil have been investigated using ^{18}O-labeled water and ^{15}N-labeled urea (Priebe and Blackmer 1989). The results suggest that preferential flow through macropores is an important factor in accounting for the fate of nitrate-nitrogen in soils of the midwest. In a practical and important test of computer groundwater model capabilities (Barcelona and Naymik 1984), it was found that monitoring data for a fertilizer contaminant plume generally supported the model prediction and demonstrated its usefulness for predicting transport and transformations of inorganic forms of nitrogen.

SUMMARY

By the mid-1980s, when these various incidents of drinking water contamination by pesticides became known by the regulatory community and the public at large, a number of questions remained unanswered. How safe is our drinking water? Are pesticides a ubiquitous contaminant of drinking water? What is the government going to do about it? What are the long-term effects of these pesticides and combinations of them in our drinking water? Many of these questions were debated in Congress, state legislatures, and in the media. While the politicians postured, scientists in industry, government, and academia were busily putting together research strategies that would address the important issues.

In order to answer these questions, there were several efforts undertaken. In a report summarizing worldwide efforts to understand the potential for groundwater contamination by pesticides (Aharonson et al. 1987), it was concluded that additional data were needed in the following five areas:

1. Chemical transformation and sorption processes in the unsaturated zone;
2. Large scale monitoring to establish a data base on the magnitude of the problem;
3. Intensive field studies to generate sufficient data to validate and calibrate models;
4. Sampling strategies to account for spatial and temporal variabilities in hydrogeology; and
5. Best management practices (BMPs) to mitigate groundwater contamination by pesticides.

As will be seen in Chapter 2, a critical part of this research strategy was the generation of statistically representative and analytically precise monitoring data concerning the true status of pesticide occurrence and concentrations in drinking water supplies. It was obvious to most scientists that the analytical chemistry and sampling procedures used in the early monitoring studies were often inadequate. Thus there was a strong push for the development of screening and confirmatory analytical techniques that would allow for the rapid, relatively nonspecific determination of whether pesticides were present in drinking water, followed by the use of highly specific techniques to confirm the presence of any suspected contaminants. It was also clear that scientifically and statistically sound sampling protocols, including the use of quality assurance and quality control techniques, would improve data quality and facilitate the training of technicians actually collecting and handling the samples.

As these monitoring data have been collected and interpreted, it has become evident that pesticide contamination of drinking water supplies can occur in a number of different ways. These mechanisms are explored further in Part 2 of this book, where they are subdivided into contamination through careless and nonagricultural uses of pesticides (Chapter 3) and contamination following the movement of properly applied materials into drinking water (Chapter 4). Much of the newer information on the environmental behavior of pesticides was initiated by the EPA in spring 1984, when the Agency required registrants of 84 different pesticides to submit new data on the physical properties of these materials relevant to the question of whether they might leach into groundwater. In addition to these data, there was an effort undertaken to identify what types of soil, climatic, and geologic conditions combined to make sites such as Long Island more susceptible to drinking water contamination.

Finally, in Part 3 the actions currently being undertaken to prevent and manage contamination of drinking water supplies by pesticides are described. In the mid-1980s, the EPA and other similar agencies around the world had not been particularly aggressive at setting appropriate maximum contaminant levels (MCLs) or health advisory levels (HALs) for all of the pesticides that had been observed in drinking water. This abrogation on the part of the Agency led to a virtual potpourri of maximum permissible concentrations in drinking water for many pesticides. For instance, in 1984 the EPA Office of Pesticide Programs within the EPA recommended a health advisory level of 30 μg/L for aldicarb, the EPA Office of Drinking Water recommended 10 μg/L, and the state of New York was utilizing 7 μg/L as an action level. As will be seen in Chapter 5, this situation has now been resolved through the Agency's promulgation of HALs and MCLs for aldicarb and many other important pesticides, but many pesticides still have no official levels established. It was obvious that some effort on the development of cost-effective treatment technologies for home-owners and publicly operated systems with contaminated drinking water was required. In the mid-1980s and still today, the most cost-effective treatment is often the use of activated carbon filters. In Chapter 5, this method and some of the other treatment systems currently available will be explored in greater depth.

The final two chapters of Part 3 describe efforts underway in the pesticide manufacturer and user community to improve the techniques for handling such materials (Chapter 6) and to develop more "environmentally friendly" pesticides (Chapter 7). It now seems likely that newer chemicals and application practices can be developed that will make the currently rare incidence of drinking water contamination even rarer.

References

Adams, C.D., and E.M. Thurman. 1991. Formation and transport of deethylatrazine in the soil and vadose zone. *J. Environ. Qual.* 20:540–47.

Aharonson, N. 1987. Potential contamination of ground water by pesticides. *Pure & Appl. Chem.* 59:1419–46.

Alhajar, B.J., G.V. Simsiman, and G. Chesters. 1990. Fate and transport of alachlor, metolachlor and atrazine in large columns. *Water Sci. Tech.* 22:87–94.

Anderson, M. 1986. Field validation of groundwater models. In *Evaluation of Pesticides in Groundwater.* American Chemical Society: Washington D.C.

Awad, T.M., W.W. Kilgore, and W. Winterlin. 1984. Movement of aldicarb in different soil types, *Bull. Environ. Contam. Toxicol.* 32:377–82.

Barcelona, M.J., and T.G. Naymik. 1984. Dynamics of a fertilizer contaminant plume in groundwater, *Environ. Sci. Technol.*, 18:257–61.

Barrett, M.R., and W.M. Williams. 1989. The occurrence of atrazine in groundwater as a result of agricultural use. Submitted for publication in the proceedings of the National Research Conference on "Pesticides in Terrestrial and Aquatic Environments," 11–12 May 1989 in Richmond, Va.

Beestman, G.B., and J.M. Deming. 1974. Dissipation of acetanilide herbicides from soils. *Agronomy Journal* 66:308–11.

Bilkert, J.N., and P.S.C. Rao. 1985. Sorption and leaching of three nonfumigant nematicides in soils. *J. Environ. Sci. Health* B20:1–26.

Buxton, H.T., T.E. Reilly, D.W. Pollock, and D.A. Smolensky. 1991. Particle tracking analysis of recharge areas on Long Island, New York. *Groundwater* 29:63–71.

Clay, S.A., and W.C. Koskinen. 1990. Adsorption and desorption of atrazine, hydroxyatrazine, and s-glutathione atrazine on two soils. *Weed Science* 38:262–66.

Cohen, S.Z., S.M. Creeger, R.F. Carsel, and C.G. Enfield. 1984. Potential pesticide contamination of groundwater from agricultural uses, in *Treatment and Disposal of Pesticide Wastes,* pp. 297–325. Washington D.C.: ACS.

Craun, G.F., D.G. Greathouse, and D.H. Gunderson. 1981. Methemoglobin levels in young children consuming high nitrate well water in the United States. *Int. J. Epidemiol.* 10:309–17.

Dahl, R., and R. Peyto. 1981. *The Causes of Cancer.* New York: Oxford University Press.

Dean, J.D., and D.F. Atwood. 1987. *Exposure Assessment: Modeling for Aldicarb in Florida.* Athens, Georgia: U.S. EPA. 4 pp.

Detroy, M.G., P.K.B. Hunt, and M.A. Holeb. 1988. Groundwater quality monitoring program in Iowa: nitrate and pesticide in shallow aquifers, in *Proceeding of the Agricultural Impacts on Ground Water, A Conference,* pp. 255–78. Dublin, Ohio: NWWA.

Eisenriech, S.J., and M.W. Sandstrom. 1987. *The Fate and Effects of Herbicides and Pesticides in Estuaries: Chemical Analysis, Report to U.S. EPA.* Minneapolis: University of Minnesota. 21 pp.

EPA. 1988. *Nitrate and Nitrite, Reviews of Environmental Contamination and Toxicology,* Vol. 107.

EPA. 1992. *Another Look: National Survey of Pesticides in Drinking Water Wells,*

Phase II Report. U.S. EPA, Office of Water and Office of Pesticides and Toxic Substances, EPA 579/09-91-020.

Gilliom, R.J., B. Alexander, and R.A. Smith. 1985. *Pesticides in the Nation's Rivers, 1975–1980, and Implications for Future Monitoring,* U.S. Geological Survey Water-Supply Paper 2271.

Good, G., and A.G. Taylor. 1987. *A Review of Agrichemical Programs and Related Water Quality Issues.* Illinois Environmental Protection Agency, Springfield, Ill.

Hersh, C.M., and W.G. Crumpton. 1989. Atrazine tolerance of algae isolated from two agricultural streams, *Environ. Toxic. Chem.* 8:327–32.

Holden, P.W. 1986. *Pesticides and Groundwater Quality.* National Academy Press: Washington, D.C. 124 pp.

Huyakorn, P.S., J.B. Kool, and T.D. Wadsworth. 1988. *A Comprehensive Simulation of Aldicarb Transport at the Wickham Site on Long Island,* Validation of Flow and Transport Models for the Unsaturated Zone Workshop, May 23–26, Ruidose, N.M.

Iowa Department of Natural Resources. 1988. *Pesticide and Synthetic Organic Chemical Survey: Report to the Iowa General Assembly on the Results of the Water System Monitoring Required by House File 2303.* Iowa Department of Natural Resources, Environmental Protection Division, Water Supply Section, Des Moines, Iowa.

Jacobs, C.A. 1988. *Domestic Well Water Sampling in Nebraska, 1987: Laboratory Findings and Their Implications.* Nebraska Department of Health, Division of Environmental Health and Housing Surveillance, Lincoln, Nebr.

Jones, R.L., A.G. Hornsby, P.S.C. Rao, and M.P. Anderson. 1987. Movement and degradation of aldicarb residues in the saturated zone under citrus groves on the Florida ridge. *J. Contam. Hydrol.* 1:265–85.

Jones, R.L., A.G. Hornsby, and P.S.C. Rao. 1988. Degradation and movement of aldicarb residues in Florida citrus soils. *Pestic. Sci.* 23:307–25.

Jones, R.L., G.W. Black, and T.L. Estes. 1986. Comparison of computer model predictions with unsaturated zone field data for aldicarb and aldoxycarb. *Environ. Toxic. Chem.* 5:1027–37.

Jones, R.L., and A.L. Bostian. 1989. Developing management practices for preventing residues of agricultural chemicals in drinking water wells. *GWMR* Fall:75–78.

Jones, R.L. 1987. Central California studies on the degradation and movement of aldicarb residues. *J. Contam. Hydrol.* 1:287–298.

Journal of American Medical Association, 7 July 1981.

Kraft, G.J., and P.A. Helmke. 1991. Aldicarb residue degradation rates at three depths of a Wisconsin central sand plain aquifer. *Pestic. Sci.* 33:47–55.

Lemley, A.T., R.J. Wagenet, and W.Z. Zhong. 1988. Sorption and degradation of aldicarb and its oxidation products in a soil-water flow system as a function of pH and temperature. *J. Environ. Qual.* 17:408–14.

Libra, R.D., G.R. Hallberg, G.G. Ressmyer, and B.E. Hoyer. 1984. Ground Water quality and hydrogeology of devonian-carbonate aquifers in Floyd and Mitchell counties, Iowa. Open File Report 84-2, Iowa Geological Survey, Iowa City, Iowa.

Lightfoot, E.N., P.S. Thorne, R.L. Jones, J.L. Hansen, and R.R. Romine. 1987. Laboratory studies on mechanisms for the degradation of aldicarb, aldicarb sulfoxide and aldicarb sulfone. *Environ. Toxic. & Chem.* 6:377–94.

Lorber, M.N., S.Z. Cohen, S.E. Noren, and G.D. Debuchananne. 1989. A national evaluation of the leaching potential of aldicarb part 1–an integrated assessment methodology. *GWMR* Fall:109–25.

Mestres, J.P., C. Spiliopoulos, R. Mestres, and J.F. Cooper. 1987. Transport of organic contaminants in groundwater: gas chromatography model to forecast the significance as applied to aldicarb sulfone residues. *Arch. Environ. Contam. Toxic.* 16:649–56.

Morgan, D.P., G.R. Hallberg, K.R. Long, R. Splinter, and L. Burmeister. 1988. An assessment of the occurrence of pesticide residues in ground water in relation to pesticide usage and agricultural practices. Progress Report, Annual Series no. 57, Iowa Pesticide Hazard Assessment Project, Department of Preventive Medicine and Environmental Health, University of Iowa, Ames, Iowa.

Milburn, P., J.E. Richards, C. Gartley, T. Pollock, H. O'Neill, and II. Bailey. 1990. Nitrate leaching from systematically tiled potato fields in New Brunswick, Canada. *J. Environ. Qual.* 19:448–54.

Ou, L.-T., K.S.V. Edvardsson, and P.S.C. Rao. 1985. Aerobic and anaerobic degradation of aldicarb in soils. *J. Agric. Food Chem.* 33:72–8.

Ou, L.-T., K.S.V. Edvardsson, J.E. Thomas, and P.S.C. Rao. 1985. Aerobic and anaerobic degradation of aldicarb sulfone in soils. *J. Agric. Food Chem.* 33:545–48.

Ou, L.-T., P.S.C. Rao, K.S.V. Edvardsson, R.E. Jessup, A.G. Hornsby and R.L. Jones. 1988. Aldicarb degradation in sandy soils from different depths. *Pestic. Sci.* 23:1–12.

Pignatello, J.J., and S.Z. Cohen. 1990. EDB regulatory review. *Rev. Environ. Contam. Tox.* 20:222–28.

Pignatello, J.J., and L.Q. Huang. 1991. Sorptive reversibility of atrazine and metolachlor residues in field soil samples. *J. Environ. Qual.* 20:222–28.

Pike, D.R., M.D. McGlamery, and E.L. Knake. 1991. A case study of herbicide use. *Weed Technology* 5:639–46.

Pothuluri, J.V., T.B. Moorman, D.C. Obenhuber, and R.D. Wauchope. 1990. Aerobic and anaerobic degradation of alachlor in samples from a surface-to-groundwater profile. *J. Environ. Qual.* 19:525–30.

Priebe, D.L., and A.M. Blackmer. 1989. Preferential movement of oxygen-18-labeled water and nitrogen-15-labeled urea through macropores in a nicollet soil. *J. Environ. Qual.* 18:66–72.

Ritter, W.F. 1990. Pesticide contamination of ground water in the United States—a review. *J. Environ. Sci. Health* B25:1–29.

Robbins, V. 1988. Pesticides in Kansas ground water. Kansas Department of Health and Environment, Bureau of Water Protection, Topeka, Kan.

Schnoor, J.L., N.B. Rao, K.J. Cartwright, and R.M. Noll. 1981. fate and transport modeling for toxic substances. In *Modeling the Fate of Chemicals in the Environment.* ed. K.L. Dickson, A.W. Maki, and J. Cairns, pp. 145–63. Ann Arbor: Ann Arbor Science.

Soileau, J.M., and R.D. Hauck. 1987. A historical view of U.S. lysimetry research with emphasis on fertilizer percolation losses. Presented at the International Conference on Infiltration Development and Application, Honolulu, Hawaii. 6–9 January 1987.

Steinberg, S.M., J.J. Pignatello, and B.L. Sawhney. 1987. Persistence of 1,2-dibromoethane in soils: entrapment in intraparticle micropores. *Environ. Sci. Technol.* 21:1201–08.

Tiedje, J.M., and M.L. Hagedorn. 1975. Degradation of alachlor by a soil fungus. *J. Agric. Food Chem.* 23:77–81.

Villeneuve, J.P., P. Lafrance, O. Banton, P. Frenchette, and C. Robert. 1988. A sensitivity analysis of adsorption and degradation parameters in the modeling of pesticide transport in soils. *J. Contam. Hydrol.* 3:77–96.

Wagenet, R.J., J.L. Hutson, and J.W. Biggar. 1989. Simulating the fate of a volatile pesticide in unsaturated soil: a case study with DBCP. *J. Environ. Qual.* 18:78–84.

Winkelmann, D.A., and S.J. Klaine. 1991a. Degradation and bound residue formation of atrazine in a western Tennessee soil. *Environ. Toxic. & Chem.* 10:335–45.

Winkelmann, D.A., and S.J. Klaine. 1991b. Degradation and bound residue formation of four atrazine metabolites, deethylatrazine, deisopropylatrazine, dealkylatrazine and hydroxyatrazine, in a western Tennessee soil. *Environ. Toxic. & Chem.* 10:347–54.

2

Review of Current Monitoring Data

As illustrated in Chapter 1, a number of relatively localized contamination episodes were identified for certain pesticides beginning in 1979. These results spawned a wide spectrum of monitoring studies that have been conducted to determine whether additional pesticides are present in the drinking water of other areas. Before reviewing the latest results, several points need to be made concerning the nature of these monitoring studies and how they should be interpreted. It's worth some time trying to describe the key elements of a well-designed monitoring study and how many of the reported studies fall far short of this ideal.

MONITORING STUDY DESIGN ISSUES

The ideal monitoring study should have four basic features:

1. A well-stated goal;
2. A statistically sound sampling program that will achieve the stated goal;
3. Documentation that all procedures of the prescribed sampling program were adhered to; and
4. Valid analytical procedures, including confirmation for all reported detections in the drinking water.

Unfortunately, many of the monitoring studies performed in the late 1970s and through the mid-1980s have not conformed to this ideal. Generally, a particular region was selected for sampling, some bottles were loaded into a van, and a few hours later the sampling was done. An analytical laboratory was identified that seemed capable of providing the analysis, and a few days later the results appeared in the local newspaper, in an attempt to create the public uproar necessary to guarantee continued funding of the "researchers"

35

involved. Even when apparently more sophisticated procedures are used to select sampling sites (Shiebe and Lettenmaier 1989), as soon as the probabilities of having selected a particular location are not estimable it becomes impossible to extrapolate results to the entire population of drinking water systems.

Study Objectives

Prior to the initiation of the monitoring program, it is necessary to have a well-stated objective or goal of the program. For instance, it might be desired to estimate, with predetermined levels of statistical power, how many people in the United States drink water containing atrazine in excess of its current MCL of 3 μg/L. This complex question would need to be answered using a multi-year sampling program, including both surface and well water collection at sampling sites throughout the use area. In the case of atrazine, this would include most of the United States. At the other end of the spectrum, the goal might be to determine whether a particular use of a pesticide (for instance DCPA (Dacthal) on golf courses) is likely to result in groundwater contamination by acidic metabolites of the compound. A narrow study such as this would be designed in an entirely different way.

Stratified Sampling Techniques

The statistical rigors of answering the questions stated in the study objective will usually necessitate the use of stratified sampling techniques in order to target sampling in a cost-effective manner. The target population is divided or stratified into subpopulations, generally using some measure of pesticide use and intrinsic susceptibility to contamination. An example of this is illustrated in Figure 2-1. In this case, it has been hypothesized that the occurrence of pesticides in surface water would be correlated with soil texture and use. Areas having soils of high clay content might be assigned a higher susceptibility to runoff than those having predominantly sandy soils. After having been assigned a susceptibility to contamination, the amount of pesticide use is estimated in order to place each surface water site into one of the nine strata. The sites within each stratum are then numbered and sampled randomly, such that each member of an individual stratum has equal chance of being selected. Certain strata may be over-sampled in order to provide greater detail about subpopulations of particular interest (for instance high-use/high-vulnerability systems), but the final results need to be properly weighted when making population-wide projections based on such uneven sampling strategies. Such stratified sampling techniques generally provide better nationwide estimates as long as the trends of contamination are not in a direction completely opposing the direction assumed when devising the stratification variable.

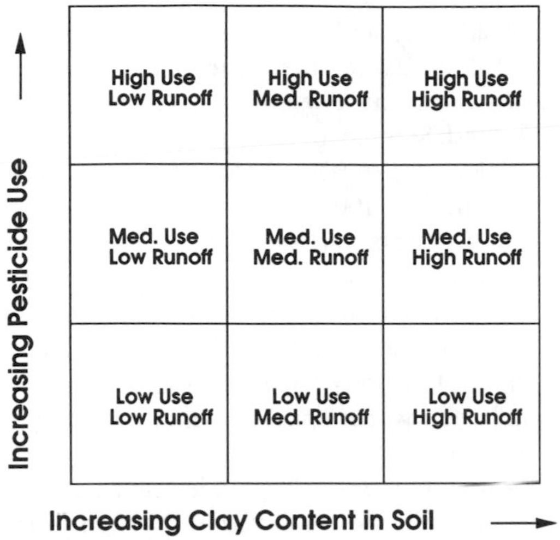

High Use Low Runoff	High Use Med. Runoff	High Use High Runoff
Med. Use Low Runoff	Med. Use Med. Runoff	Med. Use High Runoff
Low Use Low Runoff	Low Use Med. Runoff	Low Use High Runoff

Increasing Pesticide Use ↑

Increasing Clay Content in Soil ⟶

FIGURE 2-1. Stratified sampling design.

In addition to being stratified by use and susceptibility to contamination, the sample sites can be stratified temporally in order to ensure that the results are not biased as a result of sampling only during particular portions of the natural hydrogeologic cycle. One way of accomplishing this is to split the entire study into four quarters, and have each portion of the sampling program occur within the designated 3-month period. As long as this is done evenly, there will be no need to further adjust the results for this temporal factor when making population-wide projections.

Sample Collection, Integrity, and Analysis

Once individual sites have been selected, procedures for the collection of samples need to be made. This would include precise specifications, such as where and how the sample is to be taken. If drinking water wells are to be sampled, it will need to be determined where the water will be collected: near the wellhead before filtration, inside the house, or at an outside tap. There should also be an indication of how long the well should be allowed to pump in order to collect a representative sample. Surface water samples can be collected at a variety of depths and distances from the water body edge, and the sampling protocol should specify exactly where the water will be taken.

Water samples themselves should generally be placed in amber glass bottles having caps lined with polytetrafluoroethylene (PTFE). Such containers minimize the potential for the loss of pesticide due to adsorption onto the container or photolysis due to exposure to light. Insulated, corrugated shipping containers should be used for shipping. All samples should be shipped overnight using a refrigerant in the shipping container. The availability of all necessary shipping and sampling materials at each of the sample collection sites should be ensured throughout the study.

Sample chain-of-custody must be maintained throughout a complex monitoring study of this kind. Sample information chain-of-custody sheets should be supplied at each sampling location with specific instructions given on the proper techniques for filling out the forms. Generally, one section of the sheet is completed by the samplers and sent with the samples to the analytical laboratory. The receiving laboratory marks a separate section of the sheet, acknowledging their receipt, and notes any irregularities in the appearance of the samples (for example, broken bottles, not sufficiently refrigerated.) This completed chain-of-custody sheet is then kept as documentation for filing with the completed raw data package once the analytical work on each sample has been completed.

In order to demonstrate that there is no in-transit contamination, quality control samples (made up using deionized or distilled water) should be sent from the sampling sites to the analytical laboratory using the same mode of shipping as is used to send the regular samples. In addition, to show that the pesticides do not degrade in-transit, fortified samples in the concentration range of interest should be made up using otherwise uncontaminated water and shipped using identical procedures.

The persistence of several pesticides in collected groundwater samples was determined by researchers in Arkansas by simply collecting initially uncontaminated samples, fortifying them with various aqueous mixtures of pesticides, and allowing them to sit for various periods of time at either 15 or 22°C (Cavalier, Lavy, and Mattice 1991). Varying rates of degradation were observed, but they were considerably slower than the known half-lives of the compounds in either surface water or surface soils, ranging overall from 196 to 1,907 days.

One of the many technical difficulties in assessing the problem of pesticides in drinking water is the issue of analytical chemistry. The water quality reference points for most of today's pesticides are measured and quoted in units of micrograms per liter, often abbreviated as $\mu g/L$ and equated (with only slight error) to parts per billion, which is often abbreviated as PPB. A concentration on 1 $\mu g/L$ or 1 PPB is often called minute, and indeed it is when thought of as one-sixth of an aspirin tablet dissolved in a 20,000 gallon railroad tank car. To a certain extent, such comparisons are as misleading as they are comforting to the user of the contaminated water supply. The simple fact is that biological systems can have significant and potentially harmful responses to concentrations of toxic substances as low as 1 $\mu g/L$, despite how insignificant such levels might seem.

The ability of analytical chemists to routinely detect and accurately quantify such levels is—at least in part—responsible for much of the attention that pesticides in drinking water has received as an environmental issue. Because of the importance of sound analytical chemistry in accurately assessing the levels of pesticides in drinking water, some attention will be paid to the techniques used.

The first requirement for valid analytical chemistry is a well-characterized, analytical reference standard. The purity of the standard should be well in excess of 99 percent whenever possible, and there should be complete documentation of its integrity. Special care must be taken to ensure that all solvents used in standard solutions, dilutions, and extractions are free of the target analytes and are of the highest purity. Large quantities of distilled, deionized water will be required for the production of quality control samples, and this is generally obtained by passing distilled water through a commercially available cartridge water purification system.

Whenever extraction techniques are used to transfer pesticides from water into an organic solvent or solid phase column, studies must be done to verify complete recovery of the material. These recoveries should be continually verified during the study, using quality control samples fortified at known levels.

A key issue in reporting analytical results of this nature is the determination of detection limits, the minimum known concentration at which the method is able to reliably distinguish between the sample water and water known to be completely free of the analyte. A related but different issue is the determination of reporting limits, the minimum known concentration at which the analytical method gives numerical results with an acceptable degree of accuracy (such as a coefficient of variation less than 30 percent). Generally speaking, detection limits are 4 to 5 times lower than reporting limits, and the analytical protocol should specify how detections of pesticides at concentrations below the reporting limit are to be handled. One technique is to report such detections as "trace," and not assign a numerical value to the concentration. Others report the estimated concentration, but in such a way so as to indicate that the reliability of the number is in doubt. This can be accomplished by using italicized type in the report for such values or by explicitly specifying the wide confidence interval around the reported number. An extensive discussion of this issue was recently given by Keith (1991a, 1991b), who discussed the importance of reporting all analytical data, as long as it is completely documented with respect to problems and limitations.

In addition to verifying the lower end of the analytical range, care must be taken to ensure that analyzed samples are within the upper end of the calibration range and that samples are diluted and rerun whenever necessary.

When water samples are being screened for the presence of a large number of analytes, they should first be analyzed using a robust and sensitive technique, such as capillary GC with electron capture detection (ECD). Samples thought to contain particular pesticides should then be confirmed using a more specific method such as combined gas chromatography-mass spectrometry (GC-MS) with selected ion monitoring.

Special Considerations in the Sampling
Surface Water

The concentrations of pesticides in surface water are known to change rapidly in response to runoff events upstream. Without prior information to the contrary, this generally means that daily samples will need to be collected in order to provide a reasonable estimate of exposure to the people drinking the water. Daily samples can be composited into weekly (or longer-term) samples for chemical analysis in order to reduce analytical costs.

Unless other data are available showing the seasonal dependence of the pesticide behavior in the surface water, sampling should continue year-round. Sampling protocols may be written allowing the cessation of sampling after a certain number of samples showing no detectable residues have been collected. For instance, if a pesticide is used during April and May only, and no detectable residues have been observed in July, August, and September, sampling could be terminated or severely reduced until the next spring.

Recent pesticide sales or use data should be used to enumerate areas of use. This information can be cross-matched with lists showing the locations of community water systems (CWS) that utilize surface water. Additional criteria can then be established to determine whether a site will be included in the target population for sampling: (1) that the CWS uses only surface water, (2) that the CWS uses surface water year round, (3) that the CWS treats its surface water, and (4) the CWS be located in a hydrologic unit (as defined by the United States Geologic Survey) where some minimum amount of the target pesticide is used.

When selecting CWS using such criteria, however, it may often be the case that many of the water treatment plants are clustered in relatively small geographic areas. In such a case, a simple random sample of all CWS would be heavily weighted toward the regions with more sites. In order to obtain a wider geographic distribution of sampling sites, it may be decided to sample only one site per hydrologic drainage unit or some other natural geographic region.

Often, in the actual conduct of surface water monitoring programs, it will be necessary to have CWS personnel collect many or all samples. In such cases, it is essential that visits be made to the managers and operators of the cooperating water treatment plants. The program's objectives and proper sample handling and storage procedures to be used at the CWS should be discussed. Particular emphasis must be placed on sample integrity.

Usually, separate raw and finished water samples are collected at each CWS in order to determine whether the existing treatment system is effective at removing the pesticide from the drinking water. Following collection, all samples should be refrigerated at the water plants and shipped to the analytical laboratory using a refrigerant.

Data Interpretation

When all of the analytical data are in and the statistical analyses associated with the study design have been completed, the data interpretation stage begins. The most controversial aspect of such analysis is the inevitable temptation to overgeneralize the results and claim that they are more broadly applicable than the study design really allows. Since all studies are finite in time, the biggest limitation is trying to make long-term predictions about future behavior based on a relatively short-lived snapshot of what the conditions were during the time of sampling. Much of this question hinges on whether pesticide transport into drinking water has attained a steady state, given current use patterns, such that results generated in 1985 are consistent with what would be seen in 1995. As will be seen in this chapter, most studies suggest seasonal contamination of surface water by pesticides, the magnitude of which varies from year to year, but which does not appear to be getting worse with time. In the case of well water contamination by pesticides, it is clear that we continue to discover more pesticides in more wells each year as a simple result of the continually increasing scrutiny of the supplies, but it is not at all clear whether a steady state exists. Computer models would indicate that, if significant residues are going to reach groundwater, they do so within a year of treatment. This result suggests that a steady state with regard to problem wells is attained quite rapidly; however, widespread low-level contamination of groundwater is predicted to occur much more slowly. The significance of such processes to human health are questionable at this time, but further research will undoubtedly shed light on this issue.

Having summarized the various issues associated with designing, conducting, and interpreting monitoring studies, several such studies from around the world are reviewed in the sections that follow.

NATIONWIDE STUDIES IN THE UNITED STATES

Two Surveys of Drinking Water Wells: NPS and NAWWS

Although they were designed, funded, and executed by two apparently disparate entities, the EPA's National Pesticide Survey (NPS) and Monsanto's National Alachlor Well Water Survey (NAWWS) were practically identical in the methods they used to identify the number of wells likely to contain residues of pesticides (EPA 1990, 1992; Holden et al. 1992). Both were able to derive nationwide predictions with estimable precision for the number of tainted wells through the use of stratified multi-stage sampling techniques, a method advocated by a contractor involved with each, Research Triangle Institute of North Carolina. Never before attempted, the demands of providing statistically valid estimates of the number of wells having residues mandated the use of stratified sampling procedures never before employed in environmental sampling programs. Up until then, virtually all sampling programs had been very limited in geographic scope, in the number of pesticides analyzed by the method, and in the statistical power of the well selection procedures. The result of these other, primarily haphazard sampling programs was that a few pesticides had been found in a few hundred wells in a few states, but there was no valid way to extrapolate these results to obtain a nationwide view of the scope of the occurrence rates of commercial pesticides in drinking water wells.

Study Design Issues
The stated goal of the EPA National Pesticide Survey was to determine, at a national level, what proportion of the nation's drinking water wells contained detectable traces of pesticides and nitrate ion. In the case of Monsanto's survey, the company had conducted in 1985 a limited well water sampling program to check for the occurrence of alachlor in wells located in several high-use counties across the United States. A few wells containing residues of alachlor were found, but the limited nature of the sampling program prevented valid inferences being drawn about the number of wells likely to contain alachlor on a nationwide basis. Because of the widespread use of alachlor and its relatively low propensity to sorb to soil, the EPA required Monsanto to conduct a nationwide sampling program designed to estimate, with prescribed statistical power, the number of rural drinking water wells containing detectable concentrations of alachlor.

The essential problem with conducting a nationwide survey of this kind was the lack of an accurate, enumerative list of the items to be sampled—in this case private, rural wells (Ganley 1989). In the case of NPS, every public and private, rural well in the country was to be included in the sampled population; in the case of NAWWS, the population was only slightly more restrictive: private wells in every county in which alachlor was sold in 1986 (over 1,800 of the 3,100 counties in the United States). With such large and disperse populations to sample, it would clearly be unreasonable to attempt to compile a list of all 6 to 10 million such wells. Instead, the stratified, multi-stage sampling technique relies on a number of distinct stages of sampling of successively smaller geographic regions until a set of manageable sampling units is obtained. At each step care is taken to ensure that the probabilities of sampling each selected unit are properly estimated and corrected as further information is collected. This allows for the translation of the results obtained for the proportion of contaminated wells in the wells actually sampled into valid, nationwide estimates of the overall proportion of wells containing residues of the selected analyte.

In both NPS and NAWWS, the first stage of sampling involved the selection of counties. Two factors were used to stratify counties: pesticide use and groundwater susceptibility. Pesticide use information for NPS came from Doanes marketing data and similar information. For NAWWS, it was simply taken from Monsanto's own sales information. Intensity of pesticide use within the county was generally determined by dividing the county's total pounds of pesticide use by the acreage of harvested crops in the county. Counties were then stratified into either the high- or low-use category according to a predetermined criterion. For instance, it might be decided that the 25 percent of the counties with the highest pounds of pesticide use per harvested acre would be placed in the high-use category.

Groundwater susceptibility in both surveys was assigned through the use of the DRASTIC screening technique at the county-wide level. DRASTIC is described more fully in Chapter 4, but it is simply a numerical rating system that utilizes regional estimates of several properties thought to influence the susceptibility of groundwater to contamination by pesticides. These properties include depth to groundwater, recharge rates, aquifer material, soil types, topography, and properties of the vadose zone (the unsaturated region between surface soils and the groundwater table). DRASTIC was originally intended to be applied only at the subcounty level, but was used in these surveys to describe county-wide vulnerability.

As discussed earlier in this chapter, one of the major technical hurdles in the design of sampling programs such as these was properly accounting for temporal or seasonal variations in the occurrence of pesticides in groundwater (Liddle et al. 1990). This problem was handled differently within NPS and NAWWS, but the temporal effects in each were relatively minor.

Results
Monsanto began work on NAWWS in 1986. Sampling of wells was initiated in 1988 and was extended over the period of one year in order to account for any possible seasonal variations in the concentrations or occurrence percentages. Besides alachlor, the water samples were analyzed for the presence of atrazine, cyanazine, metolachlor, simazine, and nitrogen as nitrate ion. The results of this intensive study suggested that less than 1 percent of the 6 million rural drinking water wells in the alachlor use area (about 1,800 U.S. counties) contain detectable levels of alachlor, despite its widespread and intensive use for nearly 20 years in these areas. Only about 1,200 wells were shown likely to contain alachlor levels in excess of its 2 μg/L maximum contaminant level

(MCL). By contrast, nearly 800,000 wells, or 12 percent of the wells included in the sampled population, were found to contain detectable (~0.01 μg/L) residues of atrazine. However, just as with alachlor, the number of wells containing atrazine above its MCL (3 μg/L) was very small—about 5,700. After atrazine, the next most commonly found herbicide was simazine, despite the fact that it was the least commonly used of the five herbicides studied. Metolachlor, an acetanilide herbicide somewhat more persistent but less commonly used than alachlor, was found at detectable levels in about the same number of wells as alachlor. The least commonly found herbicide was cyanazine, which undergoes chemical hydrolysis throughout the soil profile, even in the absence of soil microbes, a feature that apparently confers a higher degree of environmental safety to the compound.

Other than the lack of significant pesticide contamination, the most significant finding in NAWWS was the widespread nitrate contamination observed. Approximately 5 percent of the 6 million private, rural wells in the alachlor use area were found to contain nitrate in excess of its 10 mμg/L MCL. Nitrate contamination appeared to be associated strongly with agricultural activities, although it was not clear that the application of nitrogenous fertilizers were always responsible.

Sampling in the NPS also began in 1988, but took two years to complete. Many more pesticides (127) were targeted for analysis in NPS than in NAWWS. Analytical sensitivity varied greatly for the pesticides included in the survey (Munch et al. 1990), with quantification limits ranging from a minimum of 0.10 μg/L for simetryn to a maximum of 57 μg/L for fenamiphos sulfone, although most were near 1 μg/L. Estimated detection limits (EDLs) were generally 3 to 5 times lower. The EDL was determined from the results of seven or more replicate analyses of reagent water sample fortified with the concentration of the analyte that would yield a chromatographic peak with a signal-to-noise ratio of approximately 5 to 1. The standard deviation of these data was then multiplied by the student's t value appropriate for a 99-percent confidence level and a standard deviation estimate with n−1 degrees of freedom. The EDL was defined as either the concentration of analyte yielding a 5 to 1 signal-to-noise ratio or the calculated concentration, whichever was greater.

The somewhat higher detection limits of the NPS relative to NAWWS were undoubtedly responsible for the lower numbers of wells reported to contain atrazine nationwide. The NPS found that fewer than 1 percent of the nation's 10 million private rural wells contained atrazine, but this was for an analytical reporting limit of 0.12 μg/L, over 10 times higher than the reporting limit obtained in NAWWS. Acid metabolite residues of DCPA were the most commonly observed metabolites in the NPS, occurring at detectable levels in 6.4 percent of the nations public drinking water wells and 2.5 percent of the nation's private rural wells. Fortunately, no wells were found to contain residues approaching the rather high health advisory level of 4,000 μg/L for this pesticide.

Physical properties and numbers of wells contaminated nationwide by the 12 pesticides detected in drinking water wells in the NPS are listed in Table 2-1. As shown in Figure 2-2, these data suggest a positive correlation between the number of wells containing a pesticide and the pesticide's GUS leaching index (see Chapter 4). The EPA (1992) stated that the NPS results provide "evidence that the GUS index is useful for predicting pesticide occurrence in well water samples." Clearly, other factors such as the quantity of the pesticide used and the manner in which it is used are also absolutely critical in determining the extent of contamination by any particular material. Thus, many pesticides with very high GUS values were not detected in the survey.

Both the NPS and NAWWS results suggest that, in general, shallower, older wells are more likely to contain either nitrate or pesticides. Surprisingly, however, the results failed to show any predictive capabilities of the DRAS-TIC model that had been used as a stratification variable. In fact, the only statistically significant relationships were negative associations between the number of contaminated wells and county or subcounty DRASTIC scores.

Surveys of Surface Water

USGS Surface Water Study
The USGS has recently launched a 5- to 10-year research plan called the Mid-Continent Initiative in which an in-depth investigation of the environmental fate of atrazine will be conducted (Cohen 1989; Ragone et al. 1989). It will constitute the most comprehensive and expansive pesticide fate study ever undertaken. Atrazine was selected as a representative pesticide because

TABLE 2-1   Pesticides Detected in the EPA National Pesticide Survey

Pesticide	Koc (L/kg)	Soil Half-life (Days)	GUS	Wells with Detected Residues
EDB	1	180	9.02	19,200
DCPA acid metabolites	4	365	8.71	270,010
Prometon	200	500	4.59	26,120
DBCP	129	180	4.26	38,770
Atrazine	70	60	3.83	72,370
Simazine	103	80	3.78	26,180
Bentazon	21	20	3.48	7,160
Lindane	1,081	400	2.51	13,100
Dinoseb	124	20	2.48	25
Alachlor	170	15	2.08	3,140
ETU	50	7	1.94	8,470
Hexachlorobenzene	14,100	2,080	-0.50	470

Source: EPA 1992.

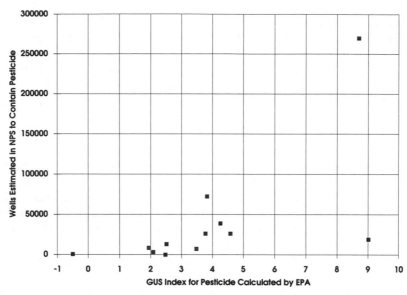

FIGURE 2-2. Correlation between pesticide occurrence in well water found in the National Pesticide Survey and the groundwater ubiquity score for the pesticide detected.

of its wide current and historic use in the United States and the large data base of analyses already in hand. The mid-continental United States was selected because of the heavy agricultural use of the basins delineated by the Missouri, Mississippi, and Ohio rivers.

The most ambitious part of the project is an attempt to calculate an overall mass balance for atrazine in the region and to develop valid models explaining the observed patterns of atrazine occurrence in drinking water. A wide range of sampling, mapping, and data analysis techniques will be required to accomplish this goal, but the USGS is better situated than any other federal agency to accomplish this, due to its existing array of water monitoring stations and mapping capabilities. The main need on the part of the USGS is an improvement in its expertise in the area of pesticide fate and transport, but this is being cultivated through extensive cooperation with both EPA and USDA researchers.

A first part of this project has been the collection of extensive amounts of surface water quality monitoring data throughout the region. The surface water sampling program is determining the extent of occurrence of atrazine and several other pesticides in the rivers, streams, and lakes of the mid-continental United States. Sampling thus far has been timed to occur contemporaneously with the seasonal runoff events just after the period of maximum pesticide use—typically May through June. Data analysis is still in its infancy, but the USGS has confirmed the occurrence of both atrazine and alachlor above their respective MCLs of 3 and 2 μg/L at certain periods of the year in some smaller, heavily farmed watersheds. However, none appear to violate the current MCLs when sampled quarterly, as provided for by the regulations (see Chapter 5).

Atrazine exceeded the EPA's drinking water MCL in 27 percent of the Mississippi River water samples collected during April, May, and June 1991 and alachlor exceeded its MCL in 4 percent of these same samples (*Food Chemical News* 1991a, 1991b). The survey included five herbicides: atrazine, alachlor, cyanazine, metolachlor, and simazine. Atrazine was detected in every sample collected during the 3-month period and occurred in the highest concentrations. Of these five, the least commonly seen was simazine, which is used much less than the other herbicides in this part of the country. Nevertheless, simazine was detected in about half of the samples, although never at levels above its MCL of 1 μg/L. EPA officials acknowledged these data, but stated that treatment to remove the herbicides would probably not be required for systems utilizing the Mississippi as a source of drinking water because MCL exceedence is based on four quarterly samples having an average concentration above the limit. Assuming that the other three quarterly samples were near zero, this would mean that a concentration of 12 μg/L for atrazine, 8 μg/L for alachlor, and 4 μg/L for simazine would have to be seen to trigger the need for treatment of the water. None of the findings in this 1991 study exceeded these concentrations.

Mass transport calculations by the USGS were also reported (Pereira and Rostad 1990; Food Chemical News 1991a). These calculations indicate that about 37% of the atrazine discharged into the Gulf of Mexico entered the river from streams draining Illinois and Iowa. The second largest source was the Missouri River basin, which contributed approximately 25% of the herbicide. The concentrations were said to begin an increase in early May and peaked near the end of the month. Highest monthly average concentrations occurred during June although they had fallen considerably by the end of that month.

Some evidence that pesticide contamination of drinking water may be less during a drought is provided by the multi-year sampling program conducted by the USGS in the Mississippi River basin during 1987–1989 (Pereira and Rostad 1990). The USGS scientist found that stream loads of the herbicides atrazine, alachlor, cyanazine, metolachlor, and simazine were much lower during the drought years of 1987 and 1988 than they were during the relatively wet year of 1989.

Monsanto Surface Water Study

The EPA issued a guidance document for the alachlor registration standard in November of 1984. The registration standard required Monsanto to conduct a monitoring study to evaluate the manner and extent of contamination of surface water with alachlor. In response to this requirement and in order to provide better data on the actual levels of alachlor and other pesticides in surface water, Monsanto conducted a pair of large-scale monitoring programs in 1985 and 1986 (Gustafson 1990). Surface water from a variety of sources, including the Great Lakes, the major rivers of the Midwest (Mississippi, Missouri, and Ohio), and smaller watersheds were sampled in 1985. Because such low levels were found in the major Midwestern rivers and the Great Lakes, the 1986 sampling was confined to smaller watersheds and designed to determine the relative importance of soil type, alachlor use, and other factors in determining the observed levels.

A variety of studies had shown that pesticide residues, including those of alachlor, were occasionally present in the surface waters of intensively farmed watersheds of the United States. While suggestive, none of the previous studies had been sufficiently broad in terms of geographical extent, temporal coverage, or number of pesticides examined to allow a complete description of the occurrence and magnitude of these residues. The Monsanto surface water projects sought to remedy the situation by conducting a 2-year, 52-watershed monitoring program in which weekly composites of daily grab samples from across the most intensively farmed areas of the United States were analyzed for the presence of several heavily used pesticides.

The Monsanto surface water monitoring data come from two separate studies—one started in the spring of 1985 and the other in the spring of 1986. Both were targeted at determining the concentrations at which alachlor occurred in drinking water derived from surface sources. In the 1985 study, raw and finished water from 24 locations were examined. In addition to alachlor, eight other pesticides were also measured. The 1986 study focused on finished water and only five herbicides, and was targeted for completion by September, based on the data collected in 1985 showing the decline of alachlor to relatively low levels (< 0.2 μg/L) by that time of year. In North America, alachlor is generally applied once per year, coinciding with planting in the spring. The sites sampling smaller watersheds are shown in Figures 2-3 and 2-4, but are identified only by code number as requested by the water systems agreeing to participate.

Table 2-2 contains a summary of the results for alachlor. For each site, two concentrations are given: the maximum weekly and the annualized mean concentration. The annualized mean concentration (AMC) is the time weighted average for the entire year. In the 1985 studies, calculation of an AMC was performed by taking a simple average of all 52 weekly concentrations measured during the year (trace levels giving negative concentrations were treated as zero in these calculations). In the 1986 studies, alachlor levels were below 0.20 μg/L at the start of the study and sampling continued until alachlor was below 0.20 μg/L for 4 consecutive

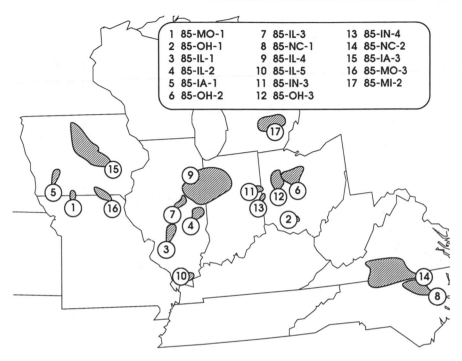

1 85-MO-1	7 85-IL-3	13 85-IN-4
2 85-OH-1	8 85-NC-1	14 85-NC-2
3 85-IL-1	9 85-IL-4	15 85-IA-3
4 85-IL-2	10 85-IL-5	16 85-MO-3
5 85-IA-1	11 85-IN-3	17 85-MI-2
6 85-OH-2	12 85-OH-3	

FIGURE 2-3.   Sampling locations from the 1985 Monsanto surface water study. Reprinted by courtesy of Marcel Dekker Inc. from Gustafson, D.I., *J. Environ. Sci. Health,* volume B25, p. 681 (1990).

weeks. Therefore, for the remaining weeks of the year for which no sampling was performed, alachlor concentrations were assumed equal to 0 μg/L, and thus represent a lower bound. The 1985 results suggest that actual alachlor AMCs in 1986 could have been, at most, as much as 0.2 μg/L higher than the values shown in Table 2-2.

Examination of the Monsanto's surface water studies shows that the occurrence of pesticides is seasonal with peak concentrations occurring, as expected, immediately following the application season. The maximum weekly concentration occurred in May or June, during the peak herbicide use season, followed by a general decline. Levels observed later in the year are principally a function of soil half-life, with the less persistent materials (cyanazine, and alachlor) occurring at low or undetectable levels. Alachlor, cyanazine, and metolachlor were never detected in any of the plants using the Great Lakes. Atrazine was detected at low levels in the plants using Lake Erie. Very low alachlor AMC's (0.01 to 0.06 $\mu\mu$g/L) were determined for those systems using major Midwestern rivers. Little, if any, difference was seen between corresponding raw and finished water

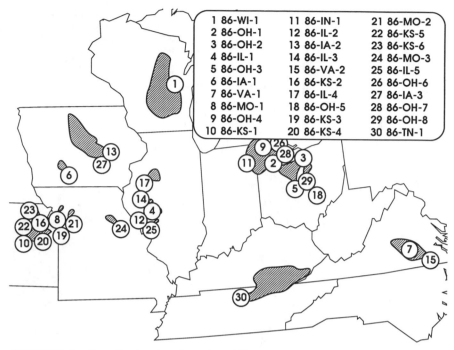

1 86-WI-1	11 86-IN-1	21 86-MO-2
2 86-OH-1	12 86-IL-2	22 86-KS-5
3 86-OH-2	13 86-IA-2	23 86-KS-6
4 86-IL-1	14 86-IL-3	24 86-MO-3
5 86-OH-3	15 86-VA-2	25 86-IL-5
6 86-IA-1	16 86-KS-2	26 86-OH-6
7 86-VA-1	17 86-IL-4	27 86-IA-3
8 86-MO-1	18 86-OH-5	28 86-OH-7
9 86-OH-4	19 86-KS-3	29 86-OH-8
10 86-KS-1	20 86-KS-4	30 86-TN-1

FIGURE 2-4.   Sampling locations from the 1986 Monsanto surface water study. Reprinted by courtesy of Marcel Dekker Inc. from Gustafson, D.I., *J. Environ. Sci. Health,* volume B25, p. 681 (1990).

from all plants sampled. This was the rationale for monitoring only finished water in the 1986 study.

Rather than using an extensive simulation to model the results, a simple multiple linear regression model was developed for predicting the annualized mean concentrations (AMC's) of alachlor, atrazine, cyanazine, and metolachlor (Gustafson 1990). Factors used as independent variables included: susceptibility of the watershed soils to runoff, physical properties of the chemical, total application rate of each chemical within the watershed, monthly precipitation totals in the watershed, and residence times of reservoirs (if any) in the watershed. Details of the collection of these variables are discussed by Gustafson (1990).

The selected regression model is summarized in Table 2-3. Herbicide use, as measured by alachlor sales and relative market-share, was the most significant variable. Following closely behind were the half-life in water, reservoir capacity, May rainfall, and K_{OC}. These five parameters were all highly significant in explaining the observed variation in AMC (p < 0.0001). Marginally less significant was soil susceptibility to runoff as

TABLE 2-2   Alachlor Concentrations in Surface Water Measured by Monsanto

Site	AMC (μg/L)[a]	MWC (μg/L)[b]	Date of Maximum
85-OH-2	1.31	10.72	6/19
85-OH-3	0.06	0.89	5/08
85-OH-1	0.17	1.26	6/26
85-IA-3	0.12	1.56	5/21
85-IA-1	0.01	<0.20	c
85-IL-2	0.01	<0.20	c
85-IL-3	0.03	0.28	5/13
85-IL-4	0.10	0.85	5/28
85-IL-5	0.01	<0.20	c
85-IL-1	0.31	4.57	6/06
85-IN-3	0.25	2.46	5/02
85-IN-4	0.55	3.49	6/12
85-MO-3	0.03	0.29	5/14
85-MI-2	0.01	<0.20	c
85-MO-1	0.01	<0.20	c
85-NC-2	0.01	<0.20	c
85-NC-1	0.03	0.26	6/19
86-OH-2	0.59	9.48	5/14
86-OH-3	0.45	3.45	6/18
86-OH-4	0.02	<0.20	c
86-OH-5	0.01	<0.20	c
86-OH-6	0.04	<0.20	c
86-OH-7	0.44	5.25	6/11
86-OH-8	0.07	1.25	6/04
86-OH-1	0.54	5.21	6/11
86-IA-2	0.43	5.07	5/21
86-IA-3	0.49	5.29	5/21
86-IA-1	0.02	<0.20	c
86-IL-2	0.01	<0.20	c
86-IL-3	0.86	5.94	5/07
86-IL-4	0.11	1.42	6/11
86-IL-5	0.75	7.44	6/11
86-IL-1	0.02	<0.20	c
86-IL-6	0.21	2.92	6/04
86-IN-1	0.53	5.03	5/21
86-KS-2	0.06	0.29	5/29
86-KS-3	0.09	0.51	5/21
86-KS-4	0.03	<0.20	c
86-KS-5	0.08	0.45	6/11
86-KS-6	0.16	0.91	7/10
86-KS-1	0.01	<0.20	c
86-MO-2	0.01	<0.20	c
86-MO-3	0.02	0.30	6/19
86-MO-1	0.11	1.12	5/14

TABLE 2-2   *(Continued)*

Site	AMC (μg/L)[a]	MWC (μg/L)[b]	Date of Maximum
86-TN-1	0.01	<0.20	c
86-VA-2	0.01	<0.20	c
86-VA-1	0.01	<0.20	c
86-WI-1	0.01	<0.20	c

Source: Gustafson 1990.
[a]Annualized mean concentration.
[b]Maximum weekly concentration.
[c]No residues above 0.20 μg/L, the reporting limit for individual analyses, were detected.

TABLE 2-3   Regression Model for Predicting Pesticide Concentrations in Surface Water

Variable	Coefficient Value	t-value	p-value
Constant	−4.7049	−12.95	<0.0001
Alachlor use (LB/A)	2.1564	7.84	<0.0001
Reservoir volume (cm)	0.15812	6.05	<0.0001
May rainfall (cm)	0.055395	5.63	<0.0001
Half-life in water (days)	0.016303	4.71	0.0001
Market-share (1.0 = alachlor)	0.6263	3.20	0.0016
Soil index (1.0–4.0, A–D)	0.24363	2.76	0.0064
Half-life in soil (days)	0.005634	2.11	0.0358
June rainfall (cm)	0.018044	1.82	0.0703

Source: Gustafson 1990.

measured by the soil index ($p = 0.0076$). April and June rainfall were not significant, nor was the half-life of the pesticide in soil.

The median absolute error of prediction with this regression is 0.0068 μg/L. In other words, half the predicted AMCs are within 0.0068 of the observed value. The relative importance of the various monthly rainfall totals is closely coupled to the time of application, most of which occurs in May. The relative importance of the pesticide physical properties suggested by the regression equation may not be widely applicable because of the rather narrow range of values spanned by the four pesticides.

UNITED KINGDOM

The European Community drinking water directive (80/778/EEC) mandates all United Kingdom water companies to supply water that meets its requirements of 0.1 μg/L per single pesticide and less than 0.5 μg/L for all pesticide residues. The United Kingdom is opposed to the European Community requirement and has repeatedly asked the Community Commission to review these standards with the aim of setting limits for individual pesticides according to their toxicity (Agrow 1991).

Nationwide Monitoring

Monitoring of UK drinking water systems during 1989–1990 indicated that the 0.1 μg/L limit was exceeded in 2.1 percent of the 540,000 analyses conducted. However, in all but 8 of the over 10,000 detections, the levels observed were below UK health advisory levels.

A total of 43 different pesticides were detected in these studies, but the most commonly seen were atrazine, simazine, chlortoluron, and mecoprop. Not surprisingly, there was considerable geographic variability in the results across the country. One system (Thames Water) had levels of atrazine and simazine above the 0.1 μg/L European Community limit in 71 percent and 34 percent of the tests, respectively.

Chief Inspector of the UK Department of the Environment's Drinking Water Inspectorate, Michael Healy, was quoted (Agrow 1991) as saying that the levels detected were "far smaller than the amounts which are known to be harmful or are likely to cause harm."

UK Granata Catchment Monitoring Study

In 1987, Great Britain began a multi-year study to measure the water quality impacts of pesticide use in the Granata catchment (watershed), which has an area of approximately 160 km^2 (Clark, Gomme, and Hennings 1991). The study includes detailed surveys of land use and pesticide applications as well as the collection of surface, ground, and rainwater samples for multi-pesticide analysis. Twenty pesticides were selected for analysis using a multi-residue method incorporating liquid-liquid extraction of the water and GC-MS quantification techniques.

The broad objectives of the study were to:

1. Determine the transport and fate of selected priority pesticides in surface and groundwater;
2. To relate measured levels of pesticides to their pattern of use and physical-chemical properties; and
3. To develop predictive models of pesticide transport and fate that can be used to develop effective management control strategies.

The pesticide and land use surveys revealed that pesticide applications in the catchment are dominated (as is true for much of Europe) by herbicides used for weed control in herbicide production: mecoprop, isoproturon, triallate, and chlortoluron.

The chemical analysis of water samples collected as part of the survey proceeded according to a rather strict protocol. Samples (10 L) were collected in bottles that had been prewashed with acid (10-percent hydrochloric), rinsed with acetone and tap water, and then with pesticide-free water from a well at the laboratory.

Extraction of all samples began within 36 hours of sample collection. At the laboratory, the sample was split in two, with a deuterated internal standard (5 μg/L) added to each subsample. The two subsamples were extracted using different procedures. The first involved a neutral or "pH ambient" extraction in which dichloromethane was used as the extraction solvent and samples were concentrated at 50°C down to 500 μL for subsequent GC-MS analysis. This method is suitable for the analysis of neutral pesticides such as atrazine.

The other subsample was submitted to acidic extraction followed by methylation. This method is suitable for the analysis of weak acids such as mecoprop. The extraction solvent was a combination of ethyl ether and dichloromethane. Once concentrated as in the other method, the extracts were methylated with diazomethane and again concentrated down to 500 μL for GC-MS analysis.

Gas chromatography with mass spectrometry as the quantification method allowed detection of the 20 pesticides down to concentrations of 0.02 to 0.13 μg/L for all but one pesticide, captafol, which had a considerably higher detection limit of 0.30 μg/L. The criterion of detection in this study was defined as the concentration above which there is a 95-percent probability that one could claim to have detected the analyte, and the limit of detection was that concentration for which there was a 5-percent probability that an analytical result was less than the criterion of detection.

Rainwater samples were collected in square stainless steel containers, with 1.165 meters sides at three sites within the watershed. The collectors were washed with acetone and pesticide-free groundwater prior to being placed at the sample collection location. The possibility of ground-contaminated splash entering the collectors was minimized by covering the surrounding ground with toughened plastic sheeting. Collected samples were filtered through a plug of silanized glass wool in order to remove insects and other large particles prior to extraction. Because at least 10 L of rainwater was required for the method, only fairly intense rainfall events (> 8 mm) could be sampled.

Surface water samples were collected at three sites along the length of the watershed's main river, the River Granata. Each site was pre-equipped with a weir and gauging station maintained by the UK National Rivers Authority. Two 5 L bottles were filled at each site from a 1 L bottle clamped to the end of a metal pole (2 m). Water was collected at mid-depth and mid-width of the river, approximately 1 meter upstream of the weir.

Groundwater samples were collected from both public supply wells and monitoring wells installed throughout the catchment. Taps at the public supply wells were allowed to run 10 minutes before the collection of samples.

Monitoring wells were pumped long enough to remove 4 times the volume of water standing in the well prior to the collection of sample.

In the case of all sampling, "blanks" were collected by transferring pesticide-free water from one container to another on-site and shipping it with the samples. Such procedures confirm that contamination could not have occurred as a result of exposure to the atmosphere or improper pre-cleaning of bottles. No instance of sample contamination was found in the Granata study.

Surface water sampling at monthly intervals along the Granata indicated fairly uniform water quality along its length with regard to the occurrence of pesticides. There was a high correlation between river stage and pesticide concentration. When the river stage was higher (later winter through early spring and in later season rain storms), pesticide concentrations were at their peak. Analysis of filtered and unfiltered surface water samples indicates that only a very small fraction of the observed pesticides are sorbed onto sediment: most of the observed material is fully dissolved. The particular pesticides observed in surface water were mainly the high-use herbicides mecoprop and isoproturon, although simazine, propyzamide, and dichlorprop were also observed. Annualized mean concentrations (AMCs) for all observed pesticides were well below 1 μg/L, which is broadly consistent with results achieved in the United States.

Groundwater samples indicated the almost complete lack of any pesticides. The only exception was the triazine herbicides, atrazine and simazine, which, despite their very low current usage in the watershed, turned up rather consistently in both monitoring wells and to a lesser extent in the public supply wells. Concentrations observed were always below 1 μg/L. The only exception to the "triazines-only rule" were two very shallow monitoring wells that had detectable levels of both isoproturon and chlortoluron, but these wells also had the two triazines present.

The Granata results suggest that most of the heavily used herbicides in this area are able to move rapidly off of fields and into surface water during periods of heavy rain, but generally not vertically into groundwater. Only the triazine herbicides, which are not as widely used, are apparently sufficiently mobile and persistent to leach via percolation processes into groundwater. This finding is entirely consistent with both the Monsanto Well Water Survey and the EPA National Pesticide Survey.

While stating that the observed levels in surface water indicate mobility, it should also be stated that the total quantity of applied pesticides reaching the major surface water body of the watershed is relatively small. Because the Granata study included use survey information, concentration determinations, and stream flow measurements, it is possible to determine the relative fraction of applied pesticides that was able to run off and appear in the surface water body. The overall proportion is less than 0.1 percent for overall pesticide use and only slightly higher than this figure for the most mobile materials. This is consistent with the results that have been obtained in other studies. For instance, in the Monsanto surface water surveys, the highest percentage ever obtained for alachlor was only 0.6 percent. These results suggest that only a small fraction of edge-of-field runoff actually reaches the major surface water bodies of regional-scale watersheds. Processes such as sedimentation and attenuation by untreated strips surrounding treated fields must remove a significant fraction of the pesticides before the major streams are reached.

The observation of only triazines in groundwater was troubling to the Granata researchers, since their pesticide use data suggested that only relatively small amounts of either simazine or atrazine are used in the watershed. They hypothesized a number of explanations for this apparent discrepancy:

1. Removal of the major agricultural pesticides by degradation or soil sorption during transport, leaving only the relatively persistent triazines;
2. A time lag in travel time for the major agricultural pesticides relative to the nonagricultural areas (railways and roadsides) where the triazine are sometimes used; and
3. Less attenuation by the soil profile in the nonagricultural areas of use posed by triazine application to road sides and railways.

The Granata authors emphasized that none of these hypotheses can be proved from their data; however, they feel quite strongly that a combination of more field data and better theoretical models of the fate and transport processes is needed in order to fully explain the results.

GERMANY

Conducted by the German pesticide industry, the IPS monitoring study included sampling of 206 wells throughout the country (IPS 1988). At 20 of these locations at least one pesticide was found above the 0.1 μg/L European Community directive. The pesticides found were atrazine, bentazon, chloridazon, simazine, pyridate, CMPP, and 1,2-dichloropropane. None of the detections were at levels in excess of 5 μg/L, and no significant health effects were suspected. The well water samples were shown to be completely free of any detectable levels of the other 28 pesticides examined.

SWEDEN

Between June 1985 and September 1987, a total of 259 monthly water samples were taken in streams from May to September and analyzed for the occurrence of 80 pesticides (Kreuger and Brink 1988). Eighteen pesticides were detected. The most frequently detected compounds were dichlorprop and MCPA with the highest concentrations found at the time of spray applications in May or June, but detectable amounts remained in surface water samples collected throughout most of the summer months. The maximum measured concentration of total phenoxy-type herbicides was 25 μg/L in one stream (or 500 times higher than European Community directive of 0.5 μg/L for total pesticide content of drinking water). Along with the phenoxy herbicides, atrazine was found in certain streams throughout the sampling period. Some streams located in areas of low agricultural acreage showed no or only sporadically detectable residues.

THE NETHERLANDS

In the first stage of research program of analyzing Dutch groundwater for residues of pesticides, the upper groundwater beneath four vulnerable soils was analyzed for nearly 2.5 years for the occurrence of 18 pesticides (Loch and Verdam, 1989). Substances found above the European Community directive of 0.1 μg/L were 1,3-dichloropropene, aldicarb, ethoprop, dinoseb, metamitron, atrazine, metolachlor, and ETU. Residues were found in groundwater only below fields having sandy soils of moderate to low organic matter content. No residues were discovered beneath a field having a clay soil with obvious macropore structures. The pesticides looked for but not found in the groundwater samples included chlorpropham, DNOC, alachlor, bentazon, captan, and amitrole.

SPAIN

In a survey of commonly used citrus herbicides in Spain (Barreda et al. 1991) it was found that atrazine, terbuthylazine, terbumeton, and bromacil were each present in nearby drinking water wells at concentrations ranging from 1 to 6 μg/L. The particular pesticides identified in these studies are consistent with findings obtained in the citrus-growing areas of the United States, but the concentrations reported are considerably less than what has been observed in Florida, which experiences much heavier rainfall.

STATE AND PROVINCIAL STUDIES IN THE UNITED STATES AND CANADA
Arkansas

In a report on the occurrence of pesticides in drinking water wells in Arkansas (Cavalier, Lavy, and Mattice 1989) summarizing sampling for the period 1985 to 1987, no detectable levels of any of the 18 targeted pesticides were found in any of the 119 sampled wells.

California

As America's biggest state in terms of agriculture, it should come as no surprise that California has conducted the most extensive water monitoring program for pesticides. A regular monitoring program involving several state agencies is now reported upon annually. The state also has some of the toughest regulations regarding the occurrence of pesticides in drinking water.

In 1988 and 1989 summaries of the data base (Cardozo et al. 1988, 1989), the ten chemicals detected in the most recent wave of sampling were 1,2-dichloropropane, DBCP, DDT, atrazine, bentazon, chlorthal-dimethyl, prometon, simazine, trifluralin, and xylene. Only the 1,2-dichloropropane, DBCP, and simazine findings were thought to be due to normal agricultural use, and the remaining pesticides found their way into the wells as a result of point sources.

In a 1990 summary of the new data added to the California well water data base (Miller et al. 1990), 14 of the 192 pesticides examined were found in at least 1 of the 2,761 new wells sampled. Only 5 of these pesticides were reported to have made their way into the groundwater as a result of normal use: aldicarb, atrazine, bromacil, diuron, and simazine.

An intensive field study of pesticide transport from the irrigated fields of the far southern Imperial Valley demonstrated that movement into groundwater was unlikely there, but that measurable concentrations of pesticides in runoff water routinely occurred (Spencer et al. 1985). Greatest runoff occurred when herbicides were applied in irrigation water. Seasonal losses of soil-applied herbicides were generally 1 to 2 percent of the amount applied. Insecticide losses were always less than 1 percent. EPTC, having been applied in the irrigation water, exhibited seasonal losses as high as 13 percent. Except for the pyrethroid insecticides, most of the pesticide transported from the field was in the water phase, as opposed to being present on the sediment.

Connecticut

A state legislative mandate in 1988 initiated the development of soil and groundwater monitoring data to evaluate whether pesticide use in the state could cause drinking water contamination (Huang and Frink 1989). Considering the proximity of Long Island within just a few miles across the narrow Long Island Sound, it is somewhat surprising that it took this long for any activity to take place. The results obtained to date do not suggest that any contaminated wells are present in the state.

Florida

Essentially a long sand bar, the state of Florida represents one of the most difficult places to use pesticides in agriculture without impacting water quality. A coordinated monitoring effort is now underway in the state, mainly following up on the many areas where EDB contamination had been discovered in the early 1980s (see Chapter 1). The monitoring program has taken on the clever nickname BADAS, standing for bromacil, atrazine, diuron, aldicarb, and simazine. These five materials have each been found in several wells at levels exceeding their respective MCLs, and the state continues to work with the individual pesticide manufacturers to remediate and manage these situations.

Georgia

In a study of groundwater quality in southeast Georgia (Beck, Asmussen, and Leonard 1985), it was found that nitrate concentrations in the exposed portions of the aquifer underlying cropland were generally higher (4 to 6 mg/L) when compared with concentrations of less than 1 mg/L in the confined portions of the aquifer and in the unconfined portions underlying forests. One well near a fertilizer storage facility had concentrations above the 10 mg/L health advisory level. In addition to simply having less nitrogen inputs in the form of fertilizer, forest lands in Georgia have been shown to have the ability to consume nitrates from shallow groundwater (Weil, Weismiller, and Turner 1990).

Iowa

The first reported sampling of Iowa drinking water for the occurrence of pesticides was performed in 1976 (Iowa DEQ 1976). In a sampling of the Iowa River from near Tama to its headwaters in late May after a series of thunderstorms, instantaneous concentrations of alachlor and atrazine exceeded 30 μg/L at one point on the river in a fairly sharp peak. Iowa is blessed with rich, thick, high organic matter soils with high water holding capacity. The intensive nature of the corn-soybean cultivation in the state makes it one of the highest in the use of pesticides like atrazine, alachlor, and others. Despite the prevalence and intensity of use over a number of years, few ground or surface water sources have significant levels of pesticide contamination. The state has conducted a number of surveys, some of which have been unique in terms of design. As pointed out in Chapter 1, all of the public water supplies, both ground and surface water, have been sampled, and the private wells in the state were sampled using a geographic stratification technique (Kelley et al. 1987).

Pesticides in the state's public systems utilizing surface water were sampled once during 1986 by the Department of Natural Resources (Iowa DNR 1988). Atrazine, alachlor, cyanazine, and metolachlor were present in nearly all of these samples, timed to coincide with the first major runoff event after May application. The Iowa DNR released results of the State Wide Rural Well Water Survey on December 13, 1989 (Iowa DNR 1990). Atrazine or one of its metabolites was detected in 8 percent of the wells, followed by metribuzin (1.9 percent), pendimethalin (1.7 percent), metolachlor (1.5 percent), cyanazine (1.2 percent), and alachlor (1.2 percent). Overall, 13.6 percent of the wells had detectable levels of at least one pesticide, but only 1.2 percent of the total wells had residues above the health advisory level or current MCL. Fully 18.3 percent of the well exceeded the 10 mg/L health advisory for nitrate-nitrogen. Shallow wells (those less than 50 feet deep) had elevated occurrence levels for both nitrates and pesticides. At the time of releasing the data, it was claimed by the study director, George Hallberg, that these were probably "best-case" results because they were collected in the midst of a major drought. However, subsequent studies by the EPA to investigate the effect of drought on the NPS results suggest that the drought did not effect results (EPA 1992).

Kansas

The Kansas Department of Health and Environment has undertaken three different studies of pesticides in groundwater (Hallberg 1989). A domestic well survey indicated that no private wells had pesticides present above the MCL, although several pesticides were detected. A separate survey of shallow public well systems revealed similar results. Finally, a third survey of a specially designed groundwater quality monitoring network revealed that only 1.6 percent had detectable levels of any pesticide.

Surface waters in Kansas have also been monitored for pesticides (Hallberg 1989). As in most other states, atrazine was the most commonly occurring material, present in 40 percent of the samples. It was followed in order of occurrence by alachlor (12 percent), metolachlor (10 percent), 2,4-D (9 percent), and metribuzin (6 percent).

Manitoba, Canada

A monitoring study was conducted during 1984 on the Ochre and Turtle Rivers of western Manitoba to determine levels of MCPA, diclofopmethyl, dicamba, bromoxynil, 2,4-D, triallate, and trifluralin (Muir and Grift 1984). All of these heavily used chemicals were detected during the sampling period, but total discharges of each represented less than 0.1 percent of the amounts estimated to be used in each watershed. The results indicate that herbicide contamination of Manitoba streams draining agricultural areas is generally low, except when major runoff occurs during the May–June application period.

Massachusetts

The Massachusetts Interagency Pesticide Task Force conducted a statewide sampling program in 1985 that focused on sites near potato fields and therefore susceptible to contamination by aldicarb and other pesticides used on that crop (Hallberg 1989). Aldicarb, 1,2-D, carbofuran, and EDB were detected. Alachlor was initially claimed to have been detected, but these analytical data have been questioned and were never duplicated.

Minnesota

In a 1986 survey of 500 vulnerable wells (Klaseus 1988), the Minnesota departments of Health and Agriculture found that traces of atrazine were present in 154 of the wells (> 30 percent of those sampled). Because of the biased sampling technique utilized, it is not known how these results would extrapolate to the remainder of the state. After atrazine, the next most commonly observed compounds were (percentage of detects in sample given in parentheses): alachlor (3.2 percent), 2,4-D (1.4 percent), metribuzin (1.2 percent), dicamba (0.8 percent), and cyanazine (0.8 percent). Several other pesticides were found in fewer than four wells: picloram, PCP, metolachlor, propachlor, MCPA, aldicarb, simazine, silvex, and EPTC. The well water samples were all found to be free of detectable levels of 15 other pesticides.

New York

Sampling for pesticides in New York has been limited primarily to Long Island. As pointed out previously in the discussion of aldicarb, a large number of wells on the island were found to contain the insecticide in 1979. Since that time, many other pesticides have been found in those wells, but generally only sporadically throughout the rest of the state. New York has been one of the most aggressive states in the country in the establishment of health advisory levels, above which some type of remediation must be practiced.

In a limited sampling program confined to other parts of the state (Walker and Porter 1989), only one very shallow well (not used for drinking) was found to have pesticide residues above the MCL (atrazine), although several pesticides were detected at levels below their respective MCLs and HALs.

North Dakota

A 3,100-hectare test site was developed by the United States Bureau of Reclamation to assess the environmental impact of interbasin transfer of Missouri River water to the James River in eastern North Dakota (Montgomery et al. 1988). In order to define baseline information, 100 observation wells were installed on a 800-meter grid and analyzed for four herbicides commonly used in the area: alachlor, atrazine, simazine, and metolachlor. Alachlor was the only material detected at the 1 μg/L detection level employed, and its presence was confirmed by repeat sampling in only one well.

Ohio

Ohio, particularly northwestern Ohio, has soils among the most sensitive to runoff in the United States. In fact, it was concern over the runoff of phosphorous from these soils into Lake Erie that prompted some of the first extensive surface water monitoring in the country. David Baker, a scientist at Heidelberg College in Tiffin, Ohio, founded the Water Quality Laboratory for just this purpose (Baker 1983a, 1983b). He first concerned himself exclusively with the levels of phosphorous in the several rivers that flow towards the lake, but he quickly realized that many pesticides were used on these lands and that they might be present in the rivers too. Operating on a small budget at first, but now having expanded to one of the most comprehensive monitoring programs in the country, the Water Quality Laboratory used state-of-the-art analytical technology combined with newly developed remote samplers to measure the occurrence of several pesticides in these rivers. He has measured surface water concentrations of atrazine, alachlor, metolachlor, and other pesticides in these areas as having annualized mean concentrations in the 1 μg/L range, but only rarely have any of the pesticides exceeded their respective MCLs.

In 1987, the Water Quality Laboratory broadened its capabilities by embarking on a cooperative private well water sampling program (Baker et al. 1989). Since that time, 80 of Ohio's 88 counties have participated in the program. In each county, local organizations such as the extension service, Farm Bureau groups, and conservation districts have sponsored a county-wide sampling program. Participating individuals picked up sample kits, filled in an information form regarding the nature of their water supply and potential local sources of contamination, collected samples on a specified date, and delivered the samples and information forms to pickup locations. As a result of a state grant, participants were charged only $1 per sample for the service.

All samples were analyzed for total nitrogen as nitrate/nitrite, ammonia, chloride, sulfate, soluble phosphorous, silica, and specific conductance. Automated procedures for this analysis were already in place as part of the Water Quality Laboratory's surface water quality monitoring programs. Pesticide analyses were conducted on samples provided by 610 program participants. Just under 3 percent of the drinking water well samples had nitrates exceeding the 10 mμg/L advisory levels. There was considerable variability in these data, with a strong correlation between observed levels of nitrates in a county and the county's DRASTIC score. There was also a strong overall correlation between well depth and the occurrence of nitrates. The predictive capabilities of DRASTIC within the state stand in stark contrast to its poor predictive capabilities reported in the nationwide surveys, NPS and NAWWS. This may reflect the ability of DRASTIC to properly rank susceptibility to contamination within a small geographic area, but not nationwide.

In selecting wells for pesticide analysis, both geographic distribution and nitrate concentrations were considered, but with such a biased sample statewide generalizations cannot be made. Of the samples analyzed, 9.2 percent contained detectable (> 0.05 μg/L) levels of atrazine, although for the reasons just mentioned there is no statistically valid way to extrapolate this figure to any meaningful statewide result. Only 0.8 percent of the samples analyzed had atrazine concentrations above 1 μg/L, and only three wells (0.5 percent of the samples analyzed) had concentrations above the 3 μg/L health advisory for atrazine. Alachlor and cyanazine were detected, but in a smaller number of wells.

Ontario, Canada

The Canadian province of Ontario has carried out several monitoring programs for the occurrence of pesticide in groundwater (Hallberg 1989). In sampling targeted only at very vulnerable areas, 95 percent of the sampled wells contained atrazine and 30 percent contained either alachlor or metolachlor. Many of these wells were subsequently determined to have been likely contaminated as a result of direct surface run-in.

Oregon

The Oregon Department of Environmental Quality coordinated an interagency project to assess the occurrence of pesticide and nitrate in the groundwater of the state (Pettit 1988). Several state and federal agencies participated in the study for which over 380 public and private drinking water wells were tested. Sampling was concentrated in those areas that were identified as being most vulnerable to contamination, as determined by factors such as pesticide use, depth to groundwater, soil properties, farming practices, and precipitation. While public water supply wells were generally free from contamination, nitrates and the pesticides EDB, DCPA, bromacil, dinoseb, aldicarb, and 1,2-dichloropropane were detected. However, only two of these detected pesticides (aldicarb and EDB) were found at levels in excess of their respective MCLs.

Pennsylvania

The herbicides alachlor, atrazine, metolachlor, and simazine have been detected in monitoring studies conducted by state personnel (Hallberg 1989). Atrazine and simazine were the most commonly found contaminants, but no concentrations above the respective MCLs have been reported for these compounds.

Waters from 20 wells in a primarily agricultural Pennsylvania watershed, the Mahantago, were analyzed for the presence of the eight most heavily used pesticides in the area: atrazine, metolachlor, cyanazine, alachlor, terbufos, chlorpyrifos, fonofos, and carbofuran (Hallberg 1989). Atrazine was found in 14 of the wells, all at concentrations below the 3 μg/L MCL, and traces of cyanazine were found in one well. None of the other pesticides were found in any sample. Wells containing > 4 mg/L nitrate ion had a higher chance of containing atrazine, although with a sample size this small the statistical power of such an observation is obviously limited.

Tennessee

Research was conducted on a 18-hectare single field watershed in western Tennessee to characterize the rate of atrazine during a 1-year period after pesticide application (Klain et al. 1988). Total loss of atrazine via runoff amounted to 1.5 percent of the amount applied, but discharge fell to undetectable levels by the time of the fourth runoff event after application. No leaching of atrazine below 20-centimeter soil depth was observed, and it was concluded that the major threat posed by the use of atrazine in the state was to surface and not groundwater supplies.

Wisconsin

Since the original aldicarb studies mentioned in Chapter 1, two major sampling efforts have been conducted in Wisconsin. The first is a problem-oriented monitoring effort directed mainly at point sources (Wisconsin DNR 1987). The other is the more comprehensive and statistically based Grade A Dairy Farm Survey (LeMasters and Doyle 1989). This study involved sampling wells at all Grade A dairy farms statewide. Approximately 10 percent of these wells were found to contain nitrates in excess of the 10 mg/L MCL, and 12 percent contained atrazine at concentrations in excess of 0.15 μg/L. These figures, when compared with the nationwide results reported in the NPS and NAWWS would suggest that the groundwater near dairy farms in Wisconsin is particularly vulnerable to contamination. This may be the result of a number of factors, including the intensity of agriculture, the shallow water table, and cool conditions leading to slower breakdown of potential contaminants at the surface.

SUMMARY OF MONITORING DATA

Taken as a whole, today's monitoring data indicate that drinking water only rarely contains pesticides at concentrations in excess of the health-based standards that have been established for each. However, there appears to be routine, seasonal occurrence of several pesticides in surface water during the peak use season as a result of runoff. There also appears to be low levels of pesticides present in many shallow, older wells. The pesticides that find their way into drinking water tend to be those that are more persistent and more mobile. No agricultural region of the world appears to have drinking water completely free of pesticide residues, but routine occurrence of levels near or above health-based guidelines occurs only in very limited geographic areas having an unfortunate combination of heavy pesticide use and high susceptibility to contamination. Further details of how this contamination takes place are explored in the next two chapters.

References

Agrow. 1991. UK drinking water quality results. *Agrow,* 16 Aug. 1991, pp. 16–17.

Baker, D.B. 1983a. *Pesticide Concentrations and Loading in Selected Lake Erie Tributaries—1982,* Final Report, U.S. EPA Grant No. R005708-01. Washington D.C.: EPA. 61 pp.

Baker, D.B. 1983b. *Studies of Sediment, Nutrient, and Pesticide Loading in Selected Lake Erie and Lake Ontario Tributaries, Part IV—Pesticide Concentrations and Loading in Selected Lake Erie Tributaries,* Draft Final Report, U.S. EPA Grant No. R005708-01, Washington D.C.: EPA. 61 pp.

Baker, D.B., L.K. Wallrabenstein, R.P. Richards, and N.L. Creamer. 1989. *Nitrates and Pesticides in Private Wells of Ohio—A State Atlas.* The Water Quality Laboratory, Heidelberg College, Tiffin, Ohio.

Barreda, D.G., E. Lorenzo M. Gamon, E. Monteagudo, A. Saez, and J.D. Cuadra. 1991. Survey of herbicide residues in soil and wells in three citrus orchards in Valencia, Spain. *Weed Research* 31:143–51.

Beck, B.F., L. Asmussen, and R. Leonard. 1985. Relationship of geology, physiography, agricultural land use, and groundwater quality in southwest Georgia. *Groundwater* 23:627–34.

Cardozo, C., C. Moore, M. Pepple, J. Troiano, and D. Weaver. 1989. *Sampling for Pesticide Residues in California Well Water, 1989 Update, Well Inventory Data Base.* Environmental Hazards Assessment Program, State of California, CDFA, Sacramento, Calif., EH 90-1. 134 pp.

Cardozo, C., M. Pepple, J. Troiano, D. Weaver, B. Fabre, S. Ali, and S. Brown. 1988. *Sampling for Pesticide Residues in California Well Water: 1988 Update, Well Inventory Data Base.* CDFA, Sacramento, Calif. 4 pp.

Cavalier, T.C., T.L. Lavy, and J.D. Mattice. 1989. Assessing Arkansas ground water for pesticides: methodology and findings. *GWMR* Fall:159–66.

Cavalier, T.C., T.L. Lavy, and J.D. Mattice. 1991. Persistence of selected pesticides in ground water samples. *Ground Water* March–April 1991:225–31.

Clark, L., G. Gomme, and S. Hennings. 1991. Study of pesticides in waters form a chalk catchment, Cambridgeshire. *Pestic. Sci.* 32:15–33.

Cohen, S.Z. 1989. Agricultural chemical news. *Ground Water Monitor Review* Fall:57–64.

EPA. 1990. *National Pesticide Survey: Project Summary.* U.S. EPA, Office of Water and Office of Pesticides and Toxic Substances. Fall 1990.

EPA. 1992. *Another Look: National Survey of Pesticides in Drinking Water Wells, Phase II Report.* U.S. EPA, Office of Water and Office of Pesticides and Toxic Substances, EPA 579/09-91-020.

Food Chemical News. 1991a. Atrazine, alachlor over MCL's in Mississippi River, USGS finds. *Pesticide & Toxic Chemical News* 20 November 1991, pp. 21–22.

Food Chemical News. 1991b. Herbicides in surface water prompt call for action. *Pesticide & Toxic Chemical News* 20 November, 1991, pp. 5–7.

Ganley, M.C. 1989. Availability and content of domestic well records in the United States. *GWMR* Fall:149–58.

Gustafson, D.I. 1990. Field calibration of surface: a model of agricultural chemicals in surface waters. *J. Environ. Sci. & Health* B25:665–87.

Hallberg, G.R. 1989. Pesticide pollution of groundwater in the humid United States. *Agriculture, Ecosystems, and Environment* 26:299–367.

Holden, L.R., J.A. Graham, R.W. Whitmore, W.J. Alexander, R.W. Pratt, S.K. Liddle, and L.L. Piper. 1992. The results of the national alachlor well water survey. *Environ. Sci. Technol.* 26:935–43.

Huang, L.Q., and C.R. Frink. 1989. Distribution of atrazine, simazine, alachlor, and metolachlor in soil profiles in Connecticut. *Bull. Environ. Contam. Toxicol.* 43:159–64.

Hubbard, R.K, and J.M. Sheridan. 1989. Nitrate movement to groundwater in the southeastern coastal plain. *J. Soil & Water Cons.* Jan–Feb:20–27.

Iowa DNR. 1988. *Pesticide and synthetic Organic Compound Survey,* Report to the Iowa General Assembly on the Results of the Water System Monitoring Required by House File 2303. Iowa DNR: Des Moines: Iowa DNR. 19 pp.

Iowa DEQ. 1976. *Water Quality Survey of the Iowa River During a Rainfall Runoff Period* #77-5. Iowa Dept. of Environ. Qual., 29 Sept. 1976.

Iowa DNR. 1990. *Iowa State-Wide Rural Well-Water Survey,* Press Releases, February 13, 1990.

IPS, 1988. *Industrieverband Pflanzenschutz Raw Water Monitoring Study.* Bonn, Germany.

Kelley, R., G.R. Hallberg, L.G. Johnson, R.D. Libra, C.A. Thompson, and R.C. Splinte. 1987. *Pesticides in Ground Water in Iowa.* Des Moines: Iowa Dept. of Water, Air, and Waste Mgmt. pp. 622–647.

Keith, L.H. 1991a. Report results right! part 1. *Chemtech* June:352–56.

Keith, L.H. 1991b. Report results right! part 2. *Chemtech* August:486–89.

Klaine, S.J., M.L. Hinman, D.A. Winklemann, K.R. Sauser, J.R. Martin, and L.W. Mo. 1988. Characterization of agricultural nonpoint pollution: pesticide migration in a west Tennessee watershed. *Environ. Toxic. Chem.* 7:609–14.

Klaseus, T. 1988. Pesticides and Groundwater: Surveys of Selected Minnesota Wells. St. Paul/Minneapolis: Minnesota Dept. of Health and Agric. 90 pp.

Kreuger, J., and N. Brink. 1988. *Losses of Pesticides from Arable Lands.* Uppsala: Swedish University of Agri. Sci. pp. 50–61.

LeMasters, G., and D.J. Doyle. 1989. *Grade A Dairy Farm Survey.* Madison, Wis.: Wisconsin Department of Agriculture, Trade, and Consumer Protection.

Liddle, S.K., R.W. Whitmore, R.E. Mason, W.J. Alexander, and L.R. Holden. 1990. Accounting for temporal variations in large-scale retrospective studies of agricultural chemicals in ground water. *GWMR* Winter 1990.

Loch, J.P.G, and B. Verdam. 1989. *Pesticide Residues in Groundwater in the Netherlands: State of Observations and Future Directions of Research.* The Hague, The Netherlands.

Miller, C., M. Pepple, J. Troiano, D. Weaver, and W. Kimaru. 1990. *Sampling for Pesticide Residues in California Well Water: 1990 Update.* Well Inventory Data Base, Fifth Annual Report to the Legislature, State Dept. of Health Services, and State Water Resources Control Board, CDFA, Sacramento, Calif. 1 December 1990. 186 pp.

Montgomery, B.R., L. Prunty, A.E. Mathison, E.C. Stegman, and W. Albus. 1988. *Nitrate and Pesticide Concentrations in Shallow Aquifers Underlying Sandy Soils,* Proceedings of the Agricultural Impacts on Ground Water: A Conference, 21–23 March 1988, at Des Moines, Iowa.

Muir, D.C.G., and N.P. Grift. 1987. Herbicide levels in rivers draining two prairie agricultural watersheds (1984). *J. Environ. Sci. Health* B22:259–84.

Munch, D.J., R.L. Graves, R.A. Maxey, and T.M. Engel. 1990. Methods development and implementation for the national pesticide survey. *Environ. Sci. Technol.* 24:1446–51.

Pereira, W.E., and C.E. Rostad. 1990. Occurrence, distributions, and transport of herbicides and their degradation products in the lower Mississippi River and its tributaries. *Environ. Sci. Technol.* 24:1400–06.

Pettit, G. 1988. Assessment of oregon's groundwater for agricultural chemicals, in *Agricultural Impacts on Ground Water—A Conference.* National Well Water Association, Des Moines, Iowa. pp. 279–95.

Ragone, S.E., M.R. Burkhart, E.M. Thurman, and C.A. Perry. 1989. *Planned Studies of Agrichemicals in Ground and Surface Water in the Mid-Continental United States.* United States Geological Survey, Reston, Va. 13 pp.

Schiebe, T.D., and D.P. Lettenmaier. 1989. Risk-based selection of monitoring wells for assessing agricultural chemical contamination of ground water. *Ground Water Monitoring Review* Fall:98–108.

South Dakota Department of Water and Natural Resources. 1989. *Pilot Sampling Program to Evaluate Pesticide and Fertilizer Impacts on Ground Water in Turner County.* Report to the 1989 South Dakota Legislature, Pierre, S.Dak.

Spencer, W.F., M.M. Cliath, J.W. Blair, and R.A. Lemert. 1985. *Transport of Pesticides from Irrigated Fields in Surface Runoff and Tile Drain Waters.* USDA ARS Conservation Report 31. 71 pp.

Walker, M.J., and K.S. Porter. 1989. *Assessment of Pesticides in Upstate New York Groundwater.* Cornell University, NY State Water Resources Institute. June 1989.

Weil, R.R., R.A. Weismiller, and R.S. Turner. 1990. Nitrate contamination of groundwater under irrigated coastal plain soils. *J. Environ. Qual.* 19:441–48.

Wisconsin DNR. 1987. *Summary of the Problem Assessment Program.* Department of Natural Resources, Madison, Wis.

II

How Does it Happen?

3

Contamination Through Carelessness and Nonagricultural Uses

Many researchers faced with the question of drinking water contamination by pesticides have focused only on the challenging technical problem of describing the behavior of the chemical after it has been applied to an agricultural field. Such uses of certain pesticides in particularly vulnerable hydrogeologic settings, conducted in full accordance with the label directions, can result in residues reaching ground or surface water supplies as a result of either leaching or runoff of the pesticide. Description of these processes will be given in Chapter 4; however, there are many other ways in which pesticides can contaminate drinking water. These mechanisms are a result of simple carelessness, neglect, and nonagricultural uses of the materials. There is still some question as to whether these events are more or less responsible than agricultural uses are for the low levels of contamination reported in the first two chapters of this book, but there is clear evidence that both are at least partially responsible (Graham 1991).

A terminology that has been used to differentiate these causes of contamination is point source vs. nonpoint source. Nonpoint sources are typically thought of as the large agricultural fields receiving relatively low concentrations of pesticides. Point sources, on the other hand, are typified by rather large concentrations of pesticides discharged to a relatively small area, such as a mixing/loading site or a storage facility. Attributing the contamination of a particular drinking water source to one or the other type of source is difficult to accomplish, but it is usually attempted. What usually occurs is that the researchers involved conduct an on-site investigation, looking for discarded pesticide containers, evidence of careless handling practices, and

FIGURE 3-1.   Potential sources of pesticides that can lead to drinking water contamination.

nearby storage facilities. The gathering of such observational evidence alone, however, does not necessarily prove that the water source has been contaminated as a result of the point source, rather than nonpoint source causes (that is, leaching from the nearest properly treated field). In order to unequivocally answer this question, it would be necessary to expend considerable effort to collect historical data and samples from new monitoring wells, and to determine local hydrogeology. Unfortunately, the lack of available resources nearly always prevents the collection of such information, and a good deal of professional judgment and simple guesswork goes into assigning the most likely source.

The remainder of this chapter will include a description of the several mechanisms other than proper agricultural use that can result in pesticide contamination of drinking water. These are the items that are generally checked in a site investigation, and they are problems that need to be corrected in regions that have shown a number of contaminated drinking water sources (see Figure 3-1).

ACCUMULATION OF SMALL SPILLS (INTENTIONAL)

It may seem obvious, but the simplest and perhaps most common way for pesticides to get into drinking water is that the material is placed directly into the well or surface water body. In the case of surface water bodies, this can occur via spills of the formulated product or spray drift during application, particularly when the material is applied by air to a field near a stream or pond. In the case of wells, the material may be spilled on the ground next to the wellhead or even directly into an open well.

During the application of pesticides, there are many times where small, seemingly insignificant spills of material occur. When accumulated over a long enough period, and if they generally occur in the same location, such spills can overload the capacity of the local soils to degrade and/or sorb the materials before off-site movement in percolating water occurs.

The first step in the process of pesticide use is the purchase and temporary storage of pesticide-containing products at the farm or other use area. While many manufacturers are moving away from the use of non-recycleable containers (see Chapter 6), these still form the vast majority of products available to the grower. Not all the containers are perfect, and, when products are stored for longer periods of time in sheds or other facilities not perfectly protected from the elements, leaks can occur. These leaking containers may not lose much material as a fraction of the amount in the bag or can, but relative to the amounts of diluted product applied in the field, these concentrations can be very significant. Such spills would have an even greater potential for causing contamination of drinking water supplies in areas where seasonal flooding is possible.

After making it through the storage step, the next opportunity for spills to occur is when it is time to mix and load the pesticide product into the application equipment. In so-called "open" systems, where individual containers are opened by hand and dumped into splashing, recirculating spray tanks, there is ample opportunity for either formulated or diluted product to find its way onto the ground. If the applicator is carrying out this operation near his or her well, which would generally be the most convenient place due to the availability of water, then such spills would have an even greater probability of resulting in the occurrence of detectable quantities of pesticide in the well.

Following the mixing procedure, it is time for the pesticide spray solution to be applied. Particularly when it is applied as a dilute aqueous spray, the user may find it necessary to discharge quantities of material in order to charge or empty lines or to simply get rid of extra material. These operations are often performed when the spray rig is stationary, resulting in relatively large doses being applied to spots in the field. Such operations can be minimized, though not entirely avoided. In so far as it is possible, these intentional discharges of additional material are now made while the spray rig is moving and away from particularly sensitive areas (that is, away from the well, ditch, or pond). The best place to do this would be on a flat, grassed buffer area alongside the treated field and far away from any drinking water sources.

After spraying, the next step in the handling operation is the cleaning of spray equipment. Just as with the mixing/loading operation, this procedure obviously requires a great deal of water, and it might be natural to perform it near the well when this is the only water source. As should be obvious, however, the great quantities of water poured onto the contaminated machinery make this a perfect way to wash pesticide residues deep into the soil profile. The better way to perform such operations is on a concrete pad where the water drains to a closed holding pond or is sufficiently diluted that significant contamination cannot occur.

The final step in the handling of the pesticide is to dispose of the containers. These can be bags, bottles, or cans. In the case of bags, they may be burned in certain areas, whereas bottles and cans should be triple rinsed (pouring the rinsate into the spray tank whenever possible) and then held in secure containers for placement into proper land fills. Most manufacturers are trying to use recycleable containers, which should minimize the need to handle materials in this way. Finally, any bags, bottles, or cans of pesticide that are not recycleable should be disposed of in a proper manner. In years past, containers have been strewn across the landscape of use areas, usually still containing residues of pesticide waiting to escape.

ACCIDENTAL MAJOR SPILLS

As the saying goes, "accidents will happen." Their probabilities for occurring may be minimized through the use of special containers and handling procedures, but with so many millions of pounds of pesticide handled year after year, it is simply inevitable that some major spills will occur. The recent spill of a railroad tank car filled with metam-sodium into the pristine upper Sacramento river brought this point home. In that case, the contamination of drinking water supplies downstream was not significant due to the dilution of the compound by the very large volume of the reservoir (Lake Shasta) and the relatively rapid rate of dissipation of the compound. However, the potential for such an incident is present every day, and precautions must be taken to ensure that particularly vulnerable water supplies are not harmed in this way.

Besides accidents during transportation, accidents can occur at manufacturing facilities, resulting in major releases of material. In the case of Bhopal, India, a pesticide intermediate was released to the atmosphere as a gas. However, such facilities handle similarly toxic materials in liquid forms that, although not resulting in the same toxic cloud, could nevertheless result in major contamination of nearby drinking water supplies. Fires at large pesticide storage facilities represent a particularly problematic form of accident that is discussed later in this chapter.

BACK-SIPHONING

Until the current practice of using so-called "mother tanks" became the recommended approach, the incidence of back-siphoning was undoubtedly one of the major causes of drinking water contamination. In the use of mother tanks, water from the well or surface water body is pumped into a tank and then driven out to the location where the water is required for the application of dilute, pesticide-containing aqueous sprays.

Before this practice, the hose from the water source would be placed directly into the application tank where the pesticides were diluted into the spray solution. If the proper check valves are not present when the pressure is turned off, it is very possible for back-siphoning of the spray solution into the water source to occur, thereby contaminating the water supply.

ABANDONED WELLS

In many parts of the world, there are no enforced regulations concerning the proper methods for closing or capping unused drinking water wells. Since they represent a direct conduit to groundwater, they are obviously a potential source of contamination. This fact has been recognized in most agricultural areas, and efforts are underway to identify such wells; however, they may be very difficult to find because of the lack of good well records in most rural regions of the world.

POORLY CONSTRUCTED WELLS

Older wells are often poorly constructed and do not meet today's standards for the integrity of the casing and methods of sealing the well. When the wellhead is not properly constructed, there is a greatly enhanced potential for runoff from the field or from mixing-loading areas to enter into the wellhead, thereby contaminating it.

More than any other single factor, the most problematic for protecting the quality of drinking water wells in the agricultural areas is the careless manner in which many of them have been constructed. A strong association between well construction quality and the likelihood of contamination has been reported in Nebraska (Exner and Spaulding 1985), and similar trends undoubtedly exist elsewhere in the world. In most areas, there are no enforced standards on well permitting and construction. Since many agricultural areas are economically depressed, the owners are pressured to install the well as cheaply as possible. This results in very shallow wells without proper casing and sealing procedures to ensure that surface water will not enter down the sides of the bore hole.

Since pesticide concentrations are generally the highest near the soil surface, the more a well is affected by surface water, the greater its chance for containing detectable residues during at least part of the year.

A properly constructed drinking water well should have the following features:

1. Optimal placement on the land surface—avoid low spots and keep away from nearby contaminant sources, such as pesticide storage areas or treated fields.
2. Its hole should be drilled to a sufficient depth, avoiding surficial aquifers whenever possible—very few wells deeper than 100 feet wind up being contaminated with pesticides at significant concentrations.
3. The bore hole should be sealed and protected with a casing that extends at least 0.15 meters (6 inches) above the land surface and higher if the well is placed in a flood-prone area.

APPLICATIONS NEAR SINKHOLES

In many areas of the world where the geologic circumstances involved the deposition of extensive calciferous materials, such as shells of crustacean origin, limestone currently forms the shallow, underlying substrate for the surface soils. Limestone is a very water soluble material and is therefore intrinsically unstable as such a soil substrate. Depending on the physical properties of the surface soils, the dissolving limestone can lead to the formation of karst topography, characterized by the occurrence of surface features known as sinkholes. These holes vary greatly in size and, as is known from even a casual acquaintance with media reports from Florida, can be large enough to consume entire homes. Dotted throughout the areas underlain by limestone are much smaller, so-called incipient karst regions, which represent a very rapid conduit for the movement of surface or near-surface soil water directly into the aquifers of such areas. Anything dissolved in the water, including pesticides, may therefore be transported directly into the groundwater with dizzying speed. As pointed out in Part 1 of this book, the practice of intensive agriculture in karstic regions such as northeastern Iowa had led to the occurrence of many pesticides in drinking water supplies, although usually at levels well below drinking water standards.

POINT SOURCES AT STORAGE FACILITIES

Sprinkled throughout the agricultural regions of today's world are numerous dealers, formulators, and distributors of products containing pesticides. As an integral part of their business, they necessarily must store significant quantities of such products. In many areas, such facilities have not taken adequate measures to protect against the inevitable leaks and broken containers. Such establishments are not immune to more catastrophic incidents (for example, fires) that can occur with particularly damaging results. Pesticide-laden containers lose their integrity in the fire, and residues can be washed deep into the soil as water is poured on the building as an avoidable part of the fire-fighting effort.

Most such buildings are not carefully placed with regard to the local hydrogeologic setting. Every effort should be made to ensure that these sites are placed such that any residues in leachate or runoff from the facilities would have sufficient time to degrade before reaching either ground or surface water sources. Data reported in Chapter 2 indicates that the potential for wells to contain detectable pesticide residues goes up rapidly whenever there is such a building within one-half mile of the well.

NONAGRICULTURAL USES OF PESTICIDES

Lawns and Golf Courses

There is some reason to believe that pesticides and fertilizers applied to home lawns and golf courses are less likely to contaminate drinking water than those used in typical agricultural situations. Grass turf and sod are very effective at reducing runoff (Cooper 1990) and, therefore, the potential for surface water contamination, and the extensive thatch layers in turf cause most organic chemicals to be removed out of any water that percolates downward (Watschke 1990). Researchers in Maryland observed that even a low density turf-grass significantly reduces sediment loss (Gross et al. 1991). Volatilization and thus overall dissipation of pesticides can be faster from turf than from soil (Cooper, Jenkins, and Curtis 1990; Stahnke et al. 1991). Turf is also a perfect host for many soil microbes that help dissipate any pesticides that might be present (BPI 1991). Researchers in Rhode Island confirmed this rapid rate of dissipation with the commonly used broad-leaf herbicides 2,4-D and dicamba (Gold et al. 1988). The degradation of isophenphos was similarly enhanced in turf-grass thatch (Niemczyk and Chapman 1987). The mobility of isazofos through a turf-grass thatch was minimal (Niemczyk and Krueger 1987). A study in Cape Cod, Massachusetts (Cohen 1990) demonstrated that seven of the test turf pesticides were never detected in nearby wells, and no currently registered pesticides were found at levels above their respective MCLs.

A somewhat contradictory result was obtained in a related study of deep nitrate movement in the unsaturated zone of a simulated urban lawn (Exner et al. 1991). In this work it was determined that as much as 95 percent of the nitrate ion applied in late August leached below the turf-grass root zone. The authors also claimed that there was sufficient nitrate in the irrigation water alone (concentration was not specified) to meet the turf-grass nitrogen needs. These results were confirmed elsewhere (Morton, Gold, and Sullivan 1988), in a study where it was shown that intentional overwatering and overfertilization will result in significant nitrate leaching. Various management strategies have been identified to help reduce these losses (Petrovic 1990). These include mainly the control of watering, soil-testing to determine actual N requirements, and controlled release formulations.

Special concerns have been raised with regard to pesticide applications to golf greens, which are constructed of very sandy soils over a gravel and tile drainage system and subjected to heavy irrigation schedules (Brown, Duble and Thomas 1977). However, the relatively small size of greens makes them relatively small point sources of nitrates or pesticides, rather than large non-point sources of the chemicals.

The net impact of the research on chemical applications to turf suggests that these uses are unlikely to result in significant drinking water contamination by pesticides due to the absorptive capacities of the thatch layer; however, nitrate losses from turf into groundwater may be a rather significant problem.

Glasshouses

Mathematical modeling and monitoring studies have been used to demonstrate that the use of several insecticides in glasshouses is unlikely to result in any significant discharges into the environment (Leistra et al. 1984). This result is consistent with what would be suggested intuitively, since a glasshouse is generally an almost completely closed system. A caveat must be included, however, for consideration of those systems that do not practice recirculation of drainage water. At some facilities, the drainage water is collected in ponds outside the glasshouse and then simply released. Because of the rather high concentrations of certain pesticides used in these operations, there is a potential for contamination if drinking water sources are located just downstream of the release point.

SUMMARY

The various potential point sources associated with drinking water contamination have been touched upon here. This brief discussion has not in any way attempted to provide the reader with an exhaustive understanding of every potential cause of drinking water contamination, but the point should have been made clear that there is more than just one way for contamination to take place: The farmer's proper application of the pesticide on an agricultural field is not always at fault.

References

BPI. 1991. Penn State studies show lawn care products do not enter ground water. *Ground Water Monitor* 4 September 1991, p. 172.

Brown, K.W., R.L. Duble, and J.C. Thomas. 1977. Influence of management and season on fate of n applied to golf greens. *Agron. J.* 69:667–71.

Cohen, S.Z. 1990. The Cape Cod study. *Golf Course Management* February:26–44.

Cooper, R.J. 1990. Evaluating the runoff and leaching potential of turfgrass pesticides. *Golf Course Management* February:9–16.

Cooper, R.J., J.J. Jenkins, and A.S. Curtis. 1990. Pendimethalin volatility following application to turfgrass. *J. Environ. Qual.* 19:508–13.

Exner, M.E., and R.F. Spaulding. 1985. Ground-water contamination and well construction in southeast Nebraska. *Ground Water* 23:26–34.

Exner, M.E., M.E. Burbach, D.G. Watts, R.C. Shearman, and R.F. Spaulding. 1991. Deep nitrate movement in the unsaturated zone of a simulated urban lawn. *J. Environ. Qual.* 20:658–62.

Gold, A.J., T.G. Morton, W.M. Sullivan, and J. McClory. 1988. Leaching of 2,4-D and dicamba from home lawns. *Water, Air, and Soil Pollut.* 37:121–29.

Graham, J.A. 1991. Monitoring groundwater and well water for crop protection chemicals. *Anal. Chem.* 63:613a–22a.

Gross, C.M., J.S. Angle, R.L. Hill, and M.S. Welterlen. 1991. Runoff and sediment losses from tall fescue under simulated rainfall. *J. Environ. Qual.* 20:604–07.

Morton, T.G., A.J. Gold, and W.M. Sullivan. 1988. Influence of overwatering and fertilization on nitrogen losses from home lawns. *J. Environ. Qual.* 17:124–30.

Niemcyzk, H.D., and H.R. Krueger. 1987. Persistence and mobility of isazofos in turfgrass thatch and soil. *J. Econ. Entomol.* 80:950–52.

Niemczyk, H.D., and R.A. Chapman. 1987. Evidence of enhanced degradation of isofenphos in turfgrass thatch and soil. *J. Econ. Entomol.* 80:880–82.

Petrovic, A.M. 1990. The fate of nitrogenous fertilizers applied to turfgrass. *J. Environ. Qual.* 19:1–14.

Stahnke, G.K., P.J. Shea, D.R. Tupy, R.N. Stougaard, and R. C. Shearman. 1991. Pendimethalin dissipation in Kentucky bluegrass turf. *Weed Science* 39:97–103.

Watschke, T.L. 1990. The environmental benefits of turfgrass and their impact on the greenhouse effect. *Golf Course Management* February:150–54.

4

Movement of Properly Applied Chemicals into Drinking Water

As seen in Chapter 3, there are a number of ways that pesticides are able to enter drinking water supplies as a result of simple carelessness or nonagricultural uses. However, the vast majority of the pesticides used worldwide are eventually applied to an agricultural field in a manner close to, if not exactly following, the label directions in terms of handling, mixing, and properly disposing of spray solutions and used containers. This chapter is concerned with the mechanisms by which these properly applied chemicals enter drinking water derived from either ground or surface sources.

The differentiation of water supplies into ground or surface sources becomes blurred when the many interactions between the two are considered. Most people would agree that a pond or stream is a surface water source and that a 200-foot-deep well is a groundwater source, but what about a hand-dug hole that is only 15 feet deep?

As this example points out, the distinction between ground and surface water is, in many ways, very artificial. The two are interconnected, and, as will be shown in Chapter 7, pesticides that are contaminants of one are nearly always also contaminants of the other. The direct intermingling between surface and groundwater can occur in several ways, the relative importance of each mechanism being dependent on the particular hydrogeologic setting of the region.

The most common place for the two to mix is at streams and rivers. During extended dry periods, larger rivers will often have a greater hydraulic head than the groundwater in the surrounding, parched soils. Darcy's law dictates that water flow will occur from the area having higher hydraulic pressure into the area having lower hydraulic pressure; thus, water will move out of the stream into the nearby soils. Such a river or stream is said to be a "losing" stream under these conditions. Contaminants, including any pesticides present in the surface water body, would be carried into the surrounding groundwater in this way.

Conversely, a "gaining" stream is one that has a lower hydraulic head than the groundwater in the surrounding soils. Such conditions usually prevail during and immediately after extended wet periods, during which time the groundwater becomes replenished. This is the more normal condition for most hydrogeologic settings, and it is usually the case that groundwater under fields is at higher hydraulic pressure than any nearby streams or rivers, which are often in very low spots topographically.

During extremely wet periods in flat areas of poor permeability, flooding can occur. Obviously, this represents a case in which groundwater is literally also surface water. As flood waters subside, the water table again assumes its normal position below the land surface, but there has usually been extensive mixing such that all surface water contaminants have had ample opportunity to blend with the groundwater.

This discussion of the mechanisms by which pesticides enter drinking water following normal use is divided into five sections. In the first, the pesticide properties affecting its potential for drinking water contamination are discussed. These include its physical-chemical properties and the methods by which the pesticide is used. Second, the environmental factors affecting the susceptibility of a particular site or region to contamination are enumerated. These include both soil and other environmental properties such as temperature and rainfall. Third, atmospheric transport is briefly discussed as it relates to the potential for longer-range transport of pesticides into drinking water than what will normally occur as a result of leaching or runoff. In the fourth section of the chapter, some of the mathematical theories that have been developed to describe pesticide fate and transport through soil are given. The use and limitations of these equations is discussed in some detail. Finally, some of the screening and simulation models describing pesticide contamination of drinking water are presented. Some of these are based on mathematical theory, and others are simply empirical relationships. Each type of model will be shown to have its place.

PESTICIDE PROPERTIES AFFECTING MOVEMENT

The physical-chemical properties of a pesticide responsible for its potential to leach into groundwater or runoff into surface water are its persistence and mobility. In many ways, mobility is the easier of the two to assess. Persistence is a compendium of many different properties, is much more sensitive to environmental factors, and is likely to be more variable. Average values for the mobility and persistence of several commercially available pesticides are given in Appendix 1.

Pesticide Mobility in Soil

The intrinsic mobility of a pesticide in soil is inversely related to its degree of sorption to (or within) soil surfaces. A pesticide having no molecular interaction with soil material would travel with the waterfront and inevitably reach groundwater unless it degraded rather rapidly. The effect of the sorptive properties on pesticide runoff have been demonstrated by experiments showing that more strongly sorbed chemicals remain nearer the soil surface than more mobile materials, leading to greater overall losses in runoff (Heathman, Ahuja, and Lehman 1985). However, the form taken by such strongly sorbed chemicals is of sediment rather than of freely dissolved chemical, making long-range transport difficult if not downright impossible.

Since water is the vector by which pesticides are moved through or over soil, via leaching and runoff water, respectively, the definitive characteristic of the pesticide molecule conferring mobility is its inherent propensity to remain dissolved in soil water as opposed to being sorbed on or in solid soil particles. This propensity is usually described by a partition coefficient, K_D (L/kg), which is equal to the ratio, at equilibrium, of the concentration of the pesticide in soil (μg/kg) to its concentration in water (μg/L).

The measurement of soil/water partition coefficients is generally conducted through a simple slurry experiment (see Figure 4-1). A known quantity of pesticide (usually radiolabeled to facilitate analysis) is added to a slurry of known water and soil content. After shaking the slurry for a set time period (24 to 48 hours), the slurry is centrifuged and the quantity of pesticide remaining in the aqueous supernatant is determined. The soil/water partition coefficient is easily calculated as the ratio of final concentrations in the soil (μg/kg) and water (μg/L) phases. There is nothing magical about waiting this period of time, but this is the conventional approach. In one kinetic study of sorption to soil (Dao and Lavy 1987), it was found that equilibration occurred within 10 to 30 minutes for four representative compounds. As will be seen below, there is now ample evidence that sorption continues even after the 1- to 2-day time period.

After this sorption step, the water is decanted off and clean water is added back into the slurry and another equilibration step is executed. In this manner the degree of reversibility in the sorption process is measured. Invariably, the partition coefficients determined in these desorption experiments are much higher than the values obtained in the original sorption experiment. This

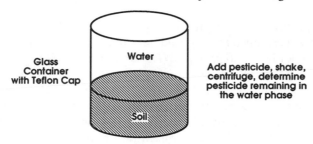

FIGURE 4-1.   Apparatus for the determination of soil sorption behavior.

phenomenon, generally known as hysteresis in the sorption/desorption process, is said to exert major influences on pesticide behavior in field situations. These hysteretic effects of sorption on the mobility of pesticides are such that major quantitative differences in leaching patterns can be found when comparing results under saturated- and unsaturated-flow conditions.

A model explaining the hysteresis in the soil sorption process has been proposed and shown to adequately describe observed behavior (Vaccari and Kaouris 1988). The model is based on the assumption that two types of sorptive sites are present in soil—one in which reversible sorption occurs and one that exhibits completely irreversible sorption. This is similar to the two-site sorption model discussed further below.

In addition to the complication of irreversibility or hysteresis, there is the issue of nonlinearity. For most pesticides in soil, the K_D ratio declines slightly as the concentration increases, and this has led to the use of more complex mathematical relationships to express the sorption behavior of the molecule. By far the most common is the Freundlich equation, an empirical relationship first used to describe the adsorption of gases onto solid surfaces, where some theoretical basis does exist (Sips 1950). To this day, much of the pesticide literature is riddled with the term, adsorption, to describe the process by which such compounds leave the soil water and become reversibly dissolved in or bound to soil particles. In reality, this is a misuse of the word, since adsorption, as defined by thermodynamicists, is a phenomenon attributed only to gases on solids. The process of pesticide equilibration between soil water and solid soil particles will be described as sorption in this book, not adsorption.

It has been pointed out that the slurry method for determining soilwater partition coefficients may be subject to experimental bias at high values for the sorption coefficient and high particle concentrations (Di Toro 1985). As subsequently explained (Mackay and Powers 1987), the effect is due to the disruption of equilibrium assumptions caused by the interaction of fast-moving particles in the shaken slurry. The practical effect is that many reported values for partition coefficients may be unrealistically low for pesticides having high values, since most were determined using the slurry method. A correction factor has been proposed (Di Toro 1985), but it requires information concerning the particle concentrations used in the slurry determinations, which may not always be available.

At sufficiently low concentrations in soil, such as those typically encountered deep within the soil profile, the sorption process becomes linear (that is, the ratio of sorbed to dissolved concentrations assumes a constant value). However, the K_D ratio for a pesticide is not constant from one soil to another. The variance in K_D is associated with many soil properties, such as soil pH, particle size distribution, and the type of clay present, but by far the most important parameter is usually the soil organic matter content, at least for surface soils containing 1 percent or more organic matter. The situation is more complex for charged molecules and those that are able to interact chemically with the clays present in most soils. The correlations also break down for aquifer materials or soils having very low amounts of organic matter and clay, in which the sorption of pesticides is significantly reduced and it becomes difficult to obtain accurate predictions of retardation factors (Stauffer, MacIntyre, and Wickman 1989). Nevertheless, most researchers currently choose to rank the relative intrinsic soil mobility of pesticides by K_{OC}, the soil/water partition coefficient divided by soil organic carbon content. As organic matter increases, a proportionate rise in K_D is generally observed. This common trend has led to the definition of K_{OC} and K_{OM}, the soilwater partition coefficient, K_D, divided by soil organic matter or soil organic carbon, respectively. The main utility of this definition is that either K_{OC} or K_{OM} may be used to predict K_D for a new soil on which the K_D of the compound has never been measured. The K_{OC} of a pesticide also serves as a soil-independent measure of the relative mobility of the pesticide in soil.

The K_{OC} of most neutral organic chemicals to soil correlates well with lipophilicity (as measured by the octanol/water partition coefficient, also known as *logP*). As discussed further in Chapter 7, *logP* can often be predicted quite well from chemical structure alone.

Recently obtained data on the kinetics of pesticide sorption to soil suggest that sorption continues to take place after the 24 to 48 hours during which equilibrium had previously been thought to have been achieved. This slow sorption process has led to the introduction of so-called two-site models to describe the process. One class of sites exhibits relatively rapid and reversible equilibrium while the other class exhibits slow but still reversible sorption. The reversibility of the soil sorption of atrazine and metolachlor in field soil samples has been studied (Pignatello and Huang 1991). The results suggest that both compounds undergo a slow sorption process under field conditions, resulting in a slow accumulation of residues in the soil. The mechanism for this was studied for a series of halogenated organic chemicals (including EDB and DBCP as class members) in which it was found that intra-particle diffusion was a likely reason for the slow attainment of equilibrium in certain matrices (Ball and Roberts 1991). These researchers confirmed that a good correlation exists between soil sorption coefficients and octanol/water partition coefficients. They also reported that the deviation from linearity in sorption coefficients was significant only at relatively high (> 50 μg/L) concentrations.

The two-site model for the nonequilibrium sorption of pesticides has been most extensively developed by workers at the University of Florida (Lee, Rao, and Brusseau 1991). The model includes the addition of two additional parameters to describe the sorption process: F, the fraction of equilibrium-type sorptive soil domains, and k_s (1/day), a first-order rate coefficient describing the slowly reversible sorption onto the portion of the sorptive soil domain not at equilibrium. Although two additional parameters are introduced, the model developers have shown that k_s and K_D for many chemicals are correlated to a high degree. Thus, the only additional independent model parameter is F, but they have observed that F is almost always very close to 0.5 and appears to be only a sluggishly increasing function of K_D.

In a dynamic situation where water is leaching downward through the soil profile, some workers have found that portions of the soil contain stagnant water that does not participate in contaminant transport. This observation has led to a further refinement in the description of pesticide sorption and transport: the so-called two-region model. A fraction of the soil water is assumed to be contained in stagnant regions not participating in transport. Pesticide is able to diffuse between the two regions, but only pesticide in the mobile region is able to leach downward with the water. Such models have also been called physical nonequilibrium models. The ability of the physical equilibrium model to predict the movement of fluometuron, aldicarb sulfone, and chloride ion in a structured soil was not adequate (Nicholls, Bromilow, and Addiscott 1982). It was found necessary to assume the existence of a fraction of soil having an immobile phase of water in order to improve the level of agreement. This is similar to what was found in attempting to predict the movement of silvex through laboratory soil columns (O'Connor, van Genuchten, and Wierenga 1976). In a study of the sorption and leaching of three nonfumigant nematicides in soils (Bilkert and Rao 1985), it was determined that aldicarb, oxamyl, and fenamiphos all leached somewhat faster through soil than would have suggested by the physical equilibrium model. The rapid rates of water movement utilized in this study may have invalidated the equilibrium assumptions. Soils with lower organic carbon contents were shown to be better described by the local equilibrium assumption models than those with higher organic carbon contents (Zurmuhl, Durner, and Herrman 1991).

It was inevitable that these two types of new sorptive/transport models would be combined into a single two-site/two-region model, as described by van Genuchten and Wagenet (1989). These researchers gave analytical solutions to such models for the special case of constant and spatially uniform water flow through soil. The practical limitations of using such models at this time include the lack of any data concerning what values to assume for the immobile fraction of soil water and mass transfer coefficients for movement between the mobile and immobile soil regions.

Another modification to these new modeling approaches was given by Boesten, van der Pas, and Smelt (1989), who proposed an even more complex three-site model for describing sorption. In a field test of this mathematical model for chemical nonequilibrium transport conducted in The Netherlands, it was found that good agreement between model predictions and observed field behavior could only be obtained when the sorption process was assumed to take place at three different classes of soil sites. Class 1 sites were assumed to attain instantaneous equilibrium; Class 2 sites equilibrated over a period of days; and Class 3 sites equilibrated over a period of months. Ignoring desorption from Class 3 sites resulted in large discrepancies between calculated and observed concentration profiles for cyanazine and metribuzin.

In addition to affecting transport of pesticides through soil, this slow sorption process should influence the rate of pesticide dissipation. This effect was recently studied in detail for three triazine herbicides: atrazine, cyanazine, and simazine (Gamerdinger, Lemley, and Wagenet 1991). The two-site model was used to describe the observed nonequilibrium sorption of the three compounds (Gamerdinger, Wagenet, and van Genuchten 1990). Sorption was found to be instantaneous for the herbicides onto 40 to 50 percent of the soil and was successfully modeled as a slower, first-order rate process onto the remaining soil fraction.

Other experiments have sought to determine what factors are responsible for the very wide range of values reported for the sorption characteristics of pesticides. In a wide-ranging complete factorial study of the mobility of 10 different pesticides and 11 of their hydrolysis metabolites in 6 different soils (Somasundaram et al. 1991), a number of important factors were confirmed. Greater levels of soil organic matter, clay, cation exchange capacity, and field capacity were each associated with lowered pesticide mobility. Increases in soil pH increased the mobility of many of these pesticides. For the group of pesticides studied, both water solubility and octanol/water partition coefficient were correlated with mobility, but octanol/water partition coefficient was a far better predictor. No direct relationship between pKa and mobility was observed.

Cationic materials are known to bond tightly to soils. The full pH behavior of the sorption of ionizable pesticides to field soils has been explored in great detail by Nicholls and Evans (1991). They found that moderately polar mono-basic weak acids were weakly sorbed in acidic soils and not sorbed at all in neutral or basic soils, where they were predominantly dissociated. Lipophilic acids are sorbed strongly even at high pH. Di-basic acids were strongly sorbed, probably by the mechanism of ligand exchange, if they were chelating agents with the potential to form 5- or 6-membered rings with an acceptor atom. Weak bases exhibited strong sorption at low pH, but sorption became relatively weak at pHs greater than 6. Zwitterions capable of chelation were very strongly sorbed at all pHs, probably due to ligand exchange interaction.

In a study of metolachlor degradation and sorption within soil (Braverman, Lavy, and Barnes 1986), it was found that metolachlor sorption to soil was positively correlated with both clay and organic carbon content. Similarly, in a study involving the sorption of nitrofen and oxyfluorfen to soil (Fadayomi and Warren 1977), it was found that both compounds were readily sorbed from solution by Ca- and H-Al-bentonite, but only slightly by Ca- and H-Al-kaolinite. In a detailed study of pirimicarb sorption by soil (Sanchez-Camazano and Sanchez-Martin 1988), a highly significant correlation between K_D and clay content was found. Sorption experiments with oxidized soil samples demonstrated that organic matter can block the sorption of pirimicarb onto clay. By contrast, chlorsulfuron was shown to bind mainly to soil organic matter and did not bind appreciably to clay (Mersie and Foy 1986).

In an exhaustive study of alachlor and atrazine sorption to 21 Korean soils, it was found that K_D was positively correlated to both clay and organic matter content (Lim, Lee, and Han 1977). Other research has confirmed the correlation of atrazine sorption and persistence with organic matter content, but the association with clay content is not always as strong.

Summarizing the current understanding of sorption processes, it is widely accepted that the linear, instantaneous equilibrium model based on K_D and K_{OC} represents an excellent first approximation. However, experimental evidence is accumulating that there are slow processes, undoubtedly related to the slow intra-particle diffusion of pesticides through the solid soil matrix, that mandate the inclusion of sorption kinetics (chemical nonequilibrium models) in any accurate description of pesticide transport through field soils. In addition, when either structured soils or soils that have rapid leaching rates are present, it becomes necessary to use two-region (physical nonequilibrium) models, having mobile and immobile soilwater phases.

Persistence in Soil

Besides mobility, the other factor determining whether a pesticide will exhibit a significant potential for contaminating drinking water is its persistence. For many years, this persistence has been described simply by the "half-life" of the compound. This terminology is a bit unfortunate for several reasons. Unlike unstable radio isotopes such as ^{14}C, which really do decay under strictly linear first-order kinetics, pesticides do not exhibit true exponential decay. In large part, this is due to the host of different processes responsible for the dissipation of pesticides in soil: volatilization, photolysis, hydrolysis, biological degradation, and oxidation, to name just a few. Each of these individual dissipative processes is in turn affected by a number of soil properties and environmental factors, such as temperature, pH, and moisture. Since soils are so variable, it follows that the degradation rate throughout the soil will also vary, sometimes by orders of magnitude. As described by Gustafson and Holden (1990), this spatial variability leads to nonlinear dissipation of pesticides in soil, despite the theoretical fact that linear first-order dissipation kinetics should be observed within a sufficiently small aliquot of the soil matrix having uniform properties.

In contrast to mobility, persistence cannot be predicted from chemical structure. Near the soil surface, the rate of field dissipation may be influenced by volatilization, photolysis, and runoff. Once it has passed below the surface, whether by incorporation or movement with rainwater, the rate of dissipation is more closely tied to the susceptibility to chemical and biological transformation reactions.

Hydrolysis and photochemical conversion of many pesticides can complement biotransformation. For example, triazines undergo chemical hydrolysis. Although photochemical conversion cannot occur below the soil surface, this is an important dissipative route for many compounds, such as foliarly-applied insecticides.

The nonlinear nature of dissipation in soil means that the concept behind describing persistence with a half-life is flawed, because the time it takes for the first 50 percent of the applied pesticide will be less than the time it takes for subsequent 50-percent portions of the remaining pesticide to dissipate (see Figure 4-2). Instead of half-life, the preferred terminology is to describe the initial dissipation of pesticides using the term DT_{50}, defined as the time necessary for the first 50 percent of the pesticide to dissipate. Similarly, DT_{90} is the time required for the first 90 percent of the pesticide to dissipate. For linear first-order kinetics, DT_{90} is just over three times longer than DT_{50}, but in real situations the ratio is generally closer to 5 or 6 due to nonlinearity in the dissipation pattern.

The dissipation pattern presented in Figure 4-2 is typical, but is not universal. The dissipation of some pesticides, particularly those under laboratory conditions, exhibits an induction period over which time the soil microbes "learn" how to degrade the pesticide. At the biomolecular level, this is probably associated with the synthesis of the appropriate degradative enzymes within the bacteria and microscopic fungi responsible for degradation. The dissipation curve in these cases will have an inflection point, as indicated in Figure 4-3. A kinetic model for pesticide biodegradation in soil having such an inflection point has been reported (Duo-Sen and Shui-Ming 1986).

In the environment, mixtures of pesticides are present at low concentrations. Some experimental evidence suggests that this leads to faster overall degradation as a result of the process of cometabolism. In a study of halogenated aliphatic compounds which would have included both EDB and DBCP as class members, a cometabolic model for biotransformation in soil was proposed and fitted to observed laboratory data (Alvarez-Cohen and

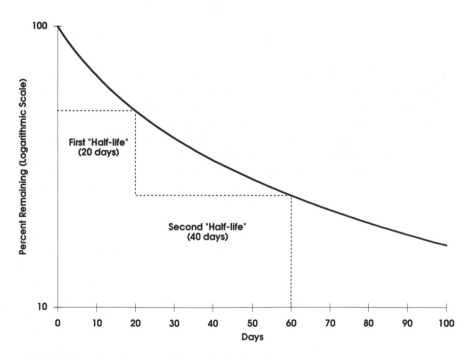

FIGURE 4-2.   Nonlinear pesticide dissipation in soil.

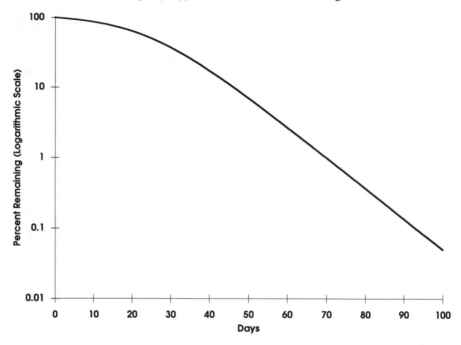

FIGURE 4-3.   Pesticide dissipation in soil exhibiting induction or lag phase.

McCarty 1991a, 1991b). At very high pesticide concentrations in soil, it has been observed that enzymatic capabilities of the microbes to degrade pesticides are depressed (Dzantro and Felsot 1990). Certain pesticides are degraded more rapidly in soils that have received previous applications of either the pesticide in question or chemically related materials (Obrigawitch et al. 1982; Harvey 1987; Harris et al. 1988; Walker and Welch 1991).

Microorganisms are essential in the conversion and detoxification of xenobiotics such as pesticides in soil (Golovleva et al. 1990). Soil enrichment techniques have been used to identify single strains of organisms capable for rapidly degrading certain classes of pesticides (Kaufman and Blake 1973). In a book chapter entitled "Kinetics of Biodegradation in Soil" (Alexander and Scow 1989), several mathematical models describing the degradation of organic pollutants by soil microbes were given. These included mathematical models based on bacterial growth, Michaelis-Menten or Monod models, first-order kinetics, compartmental models, and empirical reaction kinetics other than first-order. Other nonlinear models for pesticide dissipation have also been reported. A two-compartment model was shown to give a good description of deltamethrin dissipation in soil (Hill and Schaalje 1985), and multiple compartment models were derived (Nose 1987) to explain dissipation kinetics of fenvalerate in soil. One technique that has been reported in order to account for temperature effects on degradation is to regress concentration or its logarithm against degree-days instead of time (Nigg and Allen 1979). The use of a 3/2-order dissipation model has been reported (Stamper, Nigg, and Allen 1979), and another alternative is to use the square root of time, instead of time itself as the independent variable (Stenstrom 1989).

Alexander and Scow concluded (1989) that "given the array of chemicals, the complexity of some environments, and the variety of microorganisms that may bring about biodegradation, it is unlikely that a single model or equation would be useful for the description of rates of biodegradation of all organic substances in all environments." However, they state there is a great practical need for such a model or group of reliable models that could be validated and verified by field tests under realistic conditions. Similar conclusions about these models have been echoed by other researchers.

A problem confronting those hoping to predict dissipation rates is the considerable influence of environmental factors on the kinetics. Some of the factors said to affect the degradation of toluene in groundwater at a hazardous waste site were temperature and the concentrations of inorganic N, K, and P (Armstrong et al. 1991). The addition of K and P enhanced the rate of mineralization twofold. Changes in pH and dissolved oxygen concentration had no effect under the conditions of the experiment. In a similar study involving trichloroethylene (Barrio-Lage, Parsons, and Lorenzo 1988) methane enhanced the rate of degradation, while sulfate inhibited the degradation and ethylene had ambiguous effects. The overall degradation of chlorimuron was very similar at moisture suction values of -0.1 and -1.5 MPa and only slightly less in air dry soil (Fuesler and Hanafey 1990). Degradation rates increased with temperature in these studies. In a study of 3,4-DCA dissipation in soil, similar influences of soil moisture and temperature were found (Galiulin et al. 1984).

It is generally found that the dissipation of pesticides is faster in the field than in laboratory studies of the same compound (Frehse and Anderson 1983). Similar data have been generated regarding the mobility or leachability of pesticides. In a recent case study of pesticide degradation and leaching (Leake et al. 1987), it was found that laboratory studies overstated the potential for benazolin-ethyl to leach through soil and therefore into groundwater. However, by properly accounting for the differences in moisture and temperature conditions between laboratory and field conditions, accurate prediction of picloram field dissipation rates and field mobility from laboratory data has been reported (Hamaker, Youngson, and Goring 1967).

In a study of the effects of temperature and soil moisture on the degradation of aldicarb and oxamyl in soil (Bromilow et al. 1980a), it was determined that decreases in each lowered the rate of degradation of both compounds.

Such effects have been incorporated in mathematical models of aldicarb fate and transport in The Netherlands (Bromilow and Leistra 1980). Ferrous ions have also been identified as a factor in the rapid degradation of oxamyl, methomyl, and aldicarb in anaerobic soils (Bromilow et al. 1986).

By far the most successful in demonstrating the wide utility of such dissipation models based on soil temperature and soil moisture has been Allan Walker at the United Kingdom's Warwick Experiment Station. The pesticides for which accurate model predictions have been demonstrated include atrazine, cyanazine, and metribuzin (Smith and Walker 1989); asulam (Smith and Walker 1977); alachlor, butachlor, metolachlor, metazachlor, and propachlor (Walker and Brown 1985); and simazine and prometryn (Walker 1976).

To summarize, these data suggest that the simple linear, first-order model for pesticide dissipation so widely utilized is not particularly accurate. Effects of spatially and temporally varying soil moisture, soil temperature, and other environmental factors invariably lead to nonlinear and otherwise quite unpredictable dissipation patterns under field conditions. There is considerable evidence available in the work of Allan Walker to suggest that some of these influences can be predicted if the rate of dissipation is well quantified in the laboratory over a range of temperature and moisture conditions. Nevertheless, the situation with respect to predicting rates of dissipation and therefore persistence of pesticides in soil is not nearly as well advanced as that for predicting pesticide mobility.

Persistence in Ground and Surface Water

As discussed in Chapter 2, most studies suggest that the rate of degradation of most pesticides in groundwater is quite slow. However, the rates of degradation in surface water tend to parallel the rates observed in surface soils. The presence of sediment in estuarine water has been shown to dramatically increase the rates of degradation of most pesticides (Walker et al. 1988). Studies on the fate and short-term persistence of the insecticide permethrin demonstrated that aquatic invertebrates, plants, and stream detritus acted as major sinks for the chemical (Sundaram 1991). Other research has shown that pesticides are able to volatilize very rapidly from natural waters on which they have been inadvertently sprayed (Maguire 1991).

Water Solubility

As pointed out in Chapter 1 in the discussion of atrazine, water solubility *per se* is not generally an indicator of pesticide contamination potential. When water solubility is very low, however, it can have a substantial effect on pesticide movement (Nicholls 1988). These researchers found that simazine was retarded relative to what might have been expected based on its equilibrium partitioning behavior, and this effect has seldom found its way into most available models of pesticide transport.

A brief review of the environmental concentrations that are typically encountered demonstrates why water solubility itself is not very predictive of water contamination potential. Typical concentrations of pesticides in soil water near the surface are on the order of 1 mg/L and less, well below the solubility limit for atrazine and other chemicals known to be contaminants of drinking water. As degradation and dissipation occur, soil water concentrations deeper below the land surface are even lower, further removing water solubility from the realm of having any relevance to environmental behavior. Thus, relatively low water solubilities on the order of even 10 mg/L are not in any way limiting to the mobility of pesticides through soil and down into groundwater.

Having said this, it should be pointed out that there are two ways in which water solubility does play some role in determining the potential for drinking water contamination. Chemicals that are extremely insoluble, having solubilities less than about 1 mg/L, are sufficiently insoluble such that initially applied concentrations are much greater than what the soil water can dissolve. This will obviously decrease the potential for movement, either down through the groundwater or laterally into surface water, simply by limiting the quantity available.

The other way in which water solubility exerts some influence on the potential for drinking water contamination is through the correlation between water solubility and the propensity for interaction or sorption to soil. While there is a strong correlation between water solubility and mobility in soil, the correlation is by no means exact, and the relationship does not hold equally well for different classes of chemicals. For neutral, organic chemicals exhibiting ideal behavior, the assumption is generally quite valid, but if there are ionizable groups present or any functional groups on the molecule, such correlations lose their predictive capability. In the case of atrazine, as with many other triazines such as simazine, the water solubility is unusually depressed due to the crystal energies within the solid form of the pure chemical.

The fact that the correlation between water solubility and mobility in soil is not perfect should not come as a big surprise. The processes involved in the equilibrium between dissolved and crystalline pesticide are very different from the processes involved in determining whether the pesticide is dissolved in soil water or is sorbed into soil. Factors such as crystal energy and other intermolecular parameters are critical in determining water solubility, whereas the fraction of chemical dissolved in soil water is generally determined by other factors, such as the degree of interaction with specific chemical functionalities of materials making up soil, such as organic matter and clays.

One aspect of environmental transport of pesticides that received a flurry of attention in the mid-1980s was the purported influence of so-called humic substances on the water solubility of pesticides. It had been alleged that such materials would raise the effective water solubility of pesticides and other toxic substances, thereby leading to wider than expected contamination of drinking water supplies. As pointed out earlier, the premise on which this assumption is based is flawed, because it is the sorption coefficient, not water solubility, that controls pesticide mobility in soil. In any event, research has now confirmed that the enhancement of water solubility is negligible, except for the least soluble materials (having solubilities well below 1 mg/L). It has also been shown that the enhancement effects obtained with natural humic substances are generally far less than those achieved for the commercially prepared humic substances (Chiou et al. 1987).

Henry's Law Constant

With very volatile compounds, including EDB and DBCP, a significant fraction of exposure to the chemical as a result of its presence in household tap water can occur via the breathing of contaminated water vapor in the shower and bathroom (McKone 1987). For DBCP, up to 80 percent of the exposure was predicted to occur via this inhalation route, with only 20 percent coming from drinking the normally assumed quantity of 2 L/day. The fraction of exposure via this route is a direct function of the Henry's Law constant of the compound. The Henry's Law constant is simply the equilibrium partition coefficient for the pesticide between air and water. For an ideal solute, it is simply equal to the ratio of vapor pressure to water solubility, expressed in equivalent concentration units. Henry's Law constant is also important in determining the amount of volatilization from soil surfaces.

Method of Use

The rate, timing, and method of application can be important factors in determining the amount of pesticide leached. Compounds applied to the foliage when a crop is actively transpiring are less likely to leach than those that are soil-incorporated before planting (all else being equal). For pesticides used in areas receiving irrigation, the timing and amounts of irrigation relative to the timing of pesticide application may prove critical. In general, those pesticides applied less often, at lower rates, and later in the growing season will be less likely to contaminate drinking water supplies as a result of leaching or runoff processes.

ENVIRONMENTAL FACTORS AFFECTING MOVEMENT

Several environmental factors are known to affect the potential for pesticides to leach from or run off of treated fields into drinking water supplies. Besides certain physical properties of the soil, chemical and biological characteristics of the soil may play an important role for certain pesticides. These include pH (as it affects pesticide mobility or persistence) and the size and vigor of the soil microbial population (when biotransformation represents an important dissipation mechanism). These variables are discussed in some detail in the sections that follow.

Physical Properties of Soil

The physical properties of soil most often associated with susceptibility to leaching are low organic matter content, low field (moisture holding) capacity, and high permeability. Conversely, the factors associated with susceptibility to runoff are high field capacity and low permeability. Water movement through surface soils is a complicated process determined by the interaction of two very different forces: gravity and surface tension. Gravity acts in only one direction, down, but gravity is generally the dominant force only during those rare occasions when surface soils are completely saturated by a heavy rain or irrigation. At all other times the dominant force controlling water movement is surface tension. As soil dries, the capillary forces binding the water to the soil grow relatively stronger, resulting in greater tension or negative matric pressure, P_h. Thus, rather than flowing through the unsaturated zone, it is better to think of the water actually being sucked through the unsaturated zone into regions of more negative matric pressure.

An exception to this generalization is when heavy rainfall occurs on a structured soil able to support macropores. In this special case, water and pesticide flow have been shown to be fairly rapid processes, not unlike urban runoff flowing along the street and down through the storm sewer. Both atrazine and alachlor have been shown, under laboratory conditions, to leach faster through soils having alfalfa roots present in the profile (Zins, Wyse, and Koskinen 1991). The roots, after they decay, are thought to serve as macropores or conduits for rapid water flow.

An excellent and proven mathematical theory, Darcy's Law, is available for the description of water flow through saturated soils, but the situation in unsaturated soils is not nearly as clear-cut. The infiltration rate of water into unsaturated soil presents unique mathematical difficulties, including numerical discontinuities (Kirby 1985).

These theoretical discontinuities apparently have a counterpart in the real world in the form of so-called preferential flow, in which rapid transport of water and solutes occurs, bypassing much of the soil matrix. In a field study quantifying such preferential flow to a tile line using tracers (Everts and Kanwar 1988), it was determined that such flow contributed less than 2 percent of the total tile outflow volume, but contributed over 20 percent of the surface-applied bromide and nitrate ions that were present in the tile flow.

Simultaneous measurement of atrazine and bromide ion movement in structured soils suggests that preferential flow may affect pesticides and totally unretained solutes differently (Gish, Helling, and Kearney 1986; Starr and Glotfelty 1990). In this field study it was found that a small fraction of the applied atrazine actually moved faster than the fastest moving fraction of the applied bromide ion. Bromide ion has often been used as a tracer in field compounds and has been shown to be a useful surrogate for predicting nitrate movement through soil (Smith and Davis 1974).

The physical properties of soil can be seen to interact with pesticide properties in complex ways. The lessons and rules presented earlier in this book that lower organic matter and sandier soils are the most susceptible to leaching may indeed be true, but it is clear that phenomena other than simple uniform leaching or runoff can occur in certain soils, and these may also lead to significant quantities of pesticides reaching drinking water supplies.

Soil pH Effects

Work on the environmental fate of the sulfonyl urea herbicides suggests that pH is a critical factor for both the persistence and mobility of these materials in soil (Blair and Martin 1988). Calculated half-lives of chlorsulfuron in soil were 33, 60, 82, and 99 days as soil pH's of 6.2, 7.1, 7.7, and 8.1, respectively. Many of the newer sulfonyl ureas are less persistent than chlorsulfuron, but an effect of increased soil pH is nearly always present. Much of this is due to the pH behavior of the first-order rate constant for chemical hydrolysis (Brown 1990). Microbial degradation appears to occur regardless of pH, albeit generally at a much slower rate than hydrolysis. The mobility of these compounds also increases with increasing soil pH, although the effects are generally not as dramatic as the impact on persistence.

In a different study involving another sulfonyl-urea, sulfometuron (Lym and Swenson 1991), the importance of pH was not corroborated. However, increases in degradation rate were observed with increasing temperature and soil moisture. The pH response of the soil sorption of imazapyr, an imidazolinone herbicide, was reported to be greater than that of sulfometuron

(Wehtje et al. 1987). Considerable movement of both compounds in high pH soils was anticipated. The soil sorption from water of another ionizable pesticide, glyphosate, has been shown to be a strong function of pH (McConnel and Hossner 1985). A model has been developed that explains the pH dependence of the soil sorption of organic acid compounds (Jafvert 1990).

Binding of atrazine to humic acids decreases with increasing pH (Kalouskova 1987). This pattern of behavior is similar to what has been observed for other ionizable pesticides, for instance picloram (Goring and Hamaker, 1971). Minimum soil sorption for this herbicide was observed in alkaline sandy loam soils low in organic matter.

Taken together, these data suggest greater mobility of many pesticides in soils of higher pH. However, a compensatory effect due to the greater rate of base hydrolysis at higher pH may offset some of the influence of this enhanced mobility, at least for certain pesticides.

Soil Temperature

In nearly all cases studied, dramatic increases in the rate of pesticide dissipation at higher soil temperatures have been reported. Besides the numerous examples already cited, Zimdahl and Clark (1982) found that the degradation rates of alachlor, metolachlor, and propachlor increased as soil temperature and soil moisture increased. Similarly, soil temperature increases continued to increase the observed degradation of metribuzin at temperatures as high as 35°C (Hyzak and Zimdahl 1974).

There are also temperature effects on sorption to soil, although not nearly as dramatic and not always in the same direction. In a study involving butachlor sorption to soil as a function of temperature (Sato, Kohnosu, and Hartwig 1987), the amount of sorption increased with temperature. The observed heat of sorption (5–7 kcal/mol) indicates that sorption is a reversible, primarily physical process not involving the formation of chemical bonds. Contradictory results have been reported by researchers in India using the chemically similar herbicide, alachlor (Sethi and Chopra 1975). They found slightly less sorption as temperatures were raised from 25 to 35°C in the soils they studied. Confounding effects on water solubility increases with increasing temperature may be partly responsible for these discrepancies.

A more complete thermodynamic model of pesticide fate in unsaturated soils incorporating such temperature effects has been proposed (Lindstrom and Piver 1985), but no applications of the model to field data have been reported. Inclusion of temperature effects on the constitutive equations describing pesticide transport was also described by Canadian researchers (Padilla et al. 1988), who reported no significant effects on aldicarb but major effects on atrazine. The differences were not due to an intrinsic change in sensitivity to temperature effects (such as a higher activation energy for degradation), but were instead said to be due to the fact that atrazine stayed near the soil surface for a longer period of time because of its lesser mobility. Temperature fluctuations are much wider nearer the soil surface, hence the greater purported influence of such effects on the more persistent and less mobile material.

Spatial Variability in Soil Properties

As already mentioned, soil is a very heterogeneous material. Properly accounting for this spatial variability in models of pesticide transport is a challenging technical problem. In a computerized Monte Carlo simulation (Amoozegard-Fard, Nielsen, and Warrick 1982), sharp differences were found in the solute concentration distribution when comparing results for a deterministic value of the soil pore water velocity vs. the results in which soil pore water velocity was a random variable. The results also indicated that the variability in the dispersion coefficient was much less important than the variability in the soil pore water velocity.

An alternative approach to the problem of modeling vertical pesticide transport through soils in the presence of spatial variability is to use a transfer function model (Jury 1982). The motivation behind the use of this model is a stated pessimism by Jury that "the many causes of spatial variability of water and solute transport (for example, variations in permeability, flow through cracks or root channels, unstable flow, presence of subsurface barriers) renders measurement of the hydraulic and retention parameters of a field soil all but impossible." Because of this perceived inability to account for the flow behavior of water through soil using a deterministic approach, a stochastic model was used by Jury to characterize a system such as soil whose internal mechanisms are unknown or unknowable. A lognormal function for the transfer function was assumed and successfully fitted to a set of leaching data for sodium bromide tracer applied to the inner 0.64 ha of an 1.44-ha field of bare Tujunga loamy sand (Jury and Stolzy 1982). A similar technique was subsequently applied for a case in which transient water flow conditions existed (Jury, Dyson, and Butters 1990).

Special methods for data analysis are required due to the spatial variability exhibited by pesticide residues in field soils (Rao and Wagenet 1985; Rao et al. 1986). In an investigation of the spatial variability of pesticide soil sorption properties (Elabd, Jury, and Cliath 1986), it was found that both batch and column methods for determining K_D and K_{OC} had coefficients of variation of 25 to 38 percent. Interestingly, batch- and column-determined partition coefficients for individual soil samples were uncorrelated but gave the same overall mean value. In a similar study involving metolachlor sorption on a Captina silt loam near Fayetteville, Arkansas (Wood et al. 1987), coefficients of variation for K_D ranged from 25 to 45 percent depending on soil horizon.

Rainfall and Irrigation

Obviously, the quantity of water dumped on the field to which pesticides have been applied will play a pivotal role in determining what fraction of the applied material will move off-site into water supplies. The regression model given in Chapter 2 showed the critical influence of rainfall on the concentrations observed in surface water supplies. Similar effects have been demonstrated in the episodes of groundwater contamination by aldicarb, in which the quantity of rainfall and irrigation have been shown to directly influence the potential for leaching to occur (Jones, Black, and Estes 1986). There is also evidence suggesting that the method used to irrigate fields can exert an influence over the amount of leaching taking place.

ATMOSPHERIC TRANSPORT

Pesticides in Fog and Rainwater

Besides entering surface water via runoff and drift, some researchers have hypothesized that entrance via the atmosphere on settled dust, raindrops, or condensed fog might be an important route. There is also the issue of direct exposure to crops and humans via the air. In order to investigate this possibility, several research groups have begun to sample various portions of the atmosphere to look for pesticides. Reports of pesticides in rainfall first began in the mid-1960s. Highest concentrations are generally found as the rainstorm begins, with concentrations decreasing appreciably after prolonged rain. Generally, pesticides are observed only during times of application.

David Baker of Heidelberg College's Water Quality Laboratory in Tiffin, Ohio, has sampled rainwater at several sites, both near and several hundred miles away from heavy use areas (Richards et al. 1987). He has reported the occurrence of several commonly used herbicides at concentrations up to 10 μg/L in rainwater. Sampling in Canada has also confirmed the presence of several chlorinated pesticides in rainwater there.

Toxaphene concentrations in rainfall over a South Carolina salt marsh were monitored from 1976 to 1978 (Harder et al. 1980). During and immediately after the summer use season, toxaphene levels near 0.5 μg/L were measured, but concentrations quickly fell to undetectable levels in subsequent rainstorm events. Rainwater sampling by Wu (1980) in the Chesapeake

Bay area showed concentrations of atrazine up to 1 μg/L in samples collected during May, but concentrations fell to undetectable levels within a few weeks. Sampling by the USDA ARS also found pesticides in rainwater in the Chesapeake Bay area (Glotfelty, Seiber and Liljedahl 1987).

Glotfelty also undertook more sophisticated sampling of air and fog to determine *in situ* values for the Henry's Law constant (Glotfelty, Majewski and Seiber 1990). He found, in sampling of fog and air near Beltsville, Maryland, and in the San Joaquin Valley of California, that several pesticides were present in detectable quantities. Concentrations in the condensed fog water were often many times higher in concentration than they should have been based on a simple application on Henry's Law to concentrations found in the atmospheric air. In other words, there was an enhanced ability of the fog water to dissolve pesticides out of the air. There are several potential explanations for this. The fog droplets contain particles that undoubtedly help to sorb the pesticides. The fog droplets, as may be easily discernible from their yellowish color upon collection, are by no means pure water, and would therefore not be expected to necessarily observe Henry's Law, which applies only to dilute solutes in pure water. Finally, the curvature of the droplets due to surface tension dictates that the actual pressure inside the droplets is greatly reduced from atmospheric, enhancing the solubility of the pesticides.

Glotfelty's findings about the Henry's Law constant are at odds with those of Ligocki, Leuenberge, and Pankow (1985), who found that scavenging of such volatilized pesticides by rain is well-described by Henry's Law. It is possible that the samples collected by Glotfelty exhibited nonideal behavior because they were collected in areas typified by more contaminated atmosphere (California, Maryland) than those collected by Ligocki and his co-workers in Oregon.

Whatever the cause of the levels seen in fog and rain, the levels observed in the samples taken thus far are nowhere near the concentrations observed in runoff water leaving the edges of treated fields. Even when considering the peak concentration on the order of 10 μg/L, the overall loads of pesticides to crops and other plant species are many orders of magnitude below what would be required to cause any biological impacts. Their contribution to the loads into surface water are very limited.

Pesticide Volatilization

The measurement of rates of pesticide volatilization in the field is a challenge and has been the source of much experimentation (Taylor et al. 1976; Majewski et al. 1990; Majewski, McChesney, and Seiber 1991). The interest in such processes is both as a means of predicting concentrations to which people breathing the contaminated air will be exposed and as a tool to help predict overall field dissipation rates.

In a study of the volatilization of organic compounds in unsaturated porous media during infiltration (Cho and Jaffe 1990), it was found that the air and water phases are generally not in equilibrium and that accurate modeling of the rate of volatile losses requires knowledge of a gas-liquid mass-transfer coefficient for volatile compounds in a porous medium. The criterion of volatility above which the transfer coefficient must be utilized was not described by these researchers, but the compound investigated, TCE, has a relatively high dimensionless Henry's Law constant of 0.392 at 24.8°C, well above the values of most commercial pesticides. In a study of less volatile materials (Glotfelty et al. 1984), it was found that peak rates of volatilization occurred overnight. This counter-intuitive result was explained as a result of faster volatilization coinciding with the formation of dew on the soil during the evening. During the day when temperatures were higher, the drying of the soil and stronger sorption at the soil surface was sufficient to overcome the counterbalancing effects of increasing vapor pressure with increasing temperature.

A simple empirical equation based on vapor pressure, K_{OC} and water solubility has been shown to predict observed rates of volatilization within a factor of 10 (Glotfelty et al. 1989). Excellent correlation with vapor pressure alone has also been reported within a fairly narrow class of chemical structures (Nash 1983).

In a study of thiocarbamate pesticide volatilization from soil (Ekler 1988), it was found that formulating agents do not significantly effect the rate of loss from soil. A reciprocal relationship was found between sorption to soil and volatilization rates, permitting calculation of one of these properties when the other was known.

MATHEMATICAL THEORY OF PESTICIDE TRANSPORT IN SOIL

As soon as pesticides were first discovered in drinking water, the first question that was asked was: "How did they get there?" As was shown in Chapter 3, there are a number of explanations for how this has occurred, but in cases of contamination following normal, registered use of the compound, the answer is that the chemical has been transported there as a result of the natural flow of water. Whenever nature is involved, it can be certain that the process is complex. The task before those hoping to develop workable models of the process is to take these complex phenomena and reduce them to a set of algorithms and equations that may be solved by hand or on a computer and bear some resemblance to reality.

In embarking on this task it would be noteworthy to remember a precept proposed by the thirteenth century philosopher, Occam, who stated that when two alternate solutions yield similarly accurate fits to the observed data, than the simpler one is most likely the correct one. This principle is now known as Occam's razor, and it has come to be part of the scientific method. A hypothesis is kept as simple as possible in order to fit observed data. As more accurate data become available and conflict with the model predictions, then the model is enhanced or made more complex in order to fit the observed data. For example, Copernicus first proposed that the planets moved around the sun in perfectly circular orbits. Data were subsequently collected and interpreted by Kepler showing that the orbits were actually elliptical in nature. More recently, Einstein proposed that planetary paths were almost unimaginably complex curves in a four-dimensional time-space continuum described by general relativity theory. Is this new theory correct? Most planetary physicists would agree today, but it seems inevitable that as more accurate data are collected there will be further modifications even to this complex theory.

Unfortunately, this modeling philosophy has been largely disregarded by several model-writers in the pesticide transport field. Certain models contain algorithms for bizarre and often inconsequential processes, such as soil particle destruction by impinging raindrops. Often, there are neither the data to calibrate such sub-models nor the data showing that these arcane sub-models need to be included in order to provide a reasonable and accurate portrayal of the phenomena involved.

In a wide-ranging overview of runoff losses for pesticides commonly used in Iowa (Baker 1979), it was determined that a better mathematically based understanding of the transfer of chemicals from soil to overland flow was required. In order to establish the relevance of these concentrations, it was also pointed out that the biological disciplines need to provide information concerning the toxic effects of time-varying concentrations, durations, and loads on both ecosystems and humans.

Another use of pesticide transport models is for sensitivity analysis. A sensitivity analysis is used to determine which model inputs have the greatest effect on the model output when changed slightly. In a sensitivity analysis performed in The Netherlands (Boesten 1991), the amount of pesticide leaching was characterized by the percentage of the dose leaching below a 1-meter depth following application of the compound to a humic sand soil cropped with maize (field corn). The percentage leached was very sensitive to sorption coefficient, Freundlich exponent, and the transformation rate. The percentage leached was moderately sensitive to weather conditions, long-term sorption equilibration, and the relationship between transformation rate and temperature (the activation energy of transformation in the language of thermodynamics). Sensitivity to the extent of plant uptake was only significant for pesticides with low sorption coefficients. Sensitivity to soil hydraulic properties was small. The effect of spring vs. autumn application was very significant for non-sorbing pesticides with short half-lives. Finally, the sensitivity to spatial variability in sorption coefficient and transformation rate was substantial when only small amounts of the pesticide were predicted to leach.

In developing the theory associated with models of pesticide transport through soils, attention will be confined to those processes and equations that have been demonstrated, through comparisons with field data, to be necessary for providing an accurate description of reality.

The volumetric rate of water movement through soil, q (m/day) is described by Darcy's law:

$$q = -K_h\frac{dP_h}{dz} \qquad (4\text{-}1)$$

in which K_h (m³/kg day) is the hydraulic conductivity, P_h (kg/m day) is the hydraulic pressure, and z (m) is the depth in the soil profile.

For the special case of complete soil saturation originally studied by Darcy, K_h is a constant for an isotropic medium. However, for the unsaturated soil near the land surface, this assumption is no longer true. The permeability of soil is very much reduced as the soil dries because of the considerable resistance posed by surface tension. It is this resistance that is, in part, responsible for the fingering nature of moisture penetration paths through soil. Since water tends to move more freely through paths already wetted, and since by simply moving through a particular region the water wets the path even more, water naturally tends to move through unsaturated soil in a pattern known as preferential flow. The most extreme case of preferential flow occurs when fissures, cracks, or holes are present. As discussed earlier in this chapter, this process has sometimes been ascribed the main responsibility for unusually rapid transport of pesticides through soils that might otherwise be viewed as not vulnerable to leaching.

The point here is that K_h is not generally a constant and that some functional dependence of the hydraulic permeability upon either moisture content or hydraulic pressure must be assumed in order to provide valid solutions to the moisture flow equations. The mathematical relationships used to describe the variation in hydraulic conductivity are discussed by Hillel (1979).

Water uptake by the crop is described (Feddes 1978) according to:

$$S = \varepsilon S_p \qquad (4\text{-}2)$$

in which S (1/day) is the actual rate of uptake, S_p (1/day) is the potential rate of uptake, and ε is a unitless reduction factor. S_p is calculated from:

$$S_p = E_p/L_r \qquad (4\text{-}3)$$

in which E_p (m/day) is the potential evaporation rate and L_r (m) is the rooting depth. More sophisticated treatment of water uptake by vegetation has been described (Gardner, Jury, and Knight 1974), but the number of parameters required by such approaches preclude their use in most models of pesticide transport in soil.

The continuity equation for water movement within the soil profile is then given by:

$$\frac{d\theta}{dt} = -\frac{dq}{dz} - S \qquad (4\text{-}4)$$

in which θ is the unitless volume fraction of water in soil and t (day) is time.

This differential equation describing the moisture content of the soil and the rate of water movement through the soil profile requires boundary conditions in order to be fully specified. The condition of saturation at the water table depth means that:

$$\theta = \theta_{sat} \text{ at } z = L_{sat} \qquad (4\text{-}5)$$

in which θ_{sat} is the unitless saturated volume fraction of water for the soil of interest and L_{sat} (m) is the water table depth.

At the soil surface, there is a boundary condition on the flux of water at the surface, q_o (m/day):

$$q_o = N - E \qquad (4\text{-}6)$$

in which N (m/day) is the volumetric flux of precipitation and irrigation and E (m/day) is the actual rate of evaporation at the soil surface.

The total concentration of pesticide in soil, C^* (kg/L) is given by:

$$C^* = \theta C + \rho_s X \qquad (4\text{-}7)$$

in which C (kg/L) is the concentration in the soil water, ρ_s (kg/L) is the dry bulk density of the soil, and X (kg/kg) is the quantity of pesticide sorbed to soil. As given here, this expression neglects any pesticide present in the vapor phase of the unsaturated soil. A generalization of this relationship to account for vapor phase transport was given by Wagenet, Hutson, and Biggar (1989).

The relative amount of pesticide in the dissolved and sorbed states is generally assumed to be in equilibrium. The equilibrium relationship most often used to describe this relationship is the Freundlich equation:

$$X = K_f C^{(1/n)} \qquad (4\text{-}8)$$

in which K_f ($L^{1/n}/kg^{-1/n}$) is the Freundlich coefficient and $1/n$ is the Freundlich exponent. In the special case of $n = 1$, this becomes a linear sorption isotherm, and K_f in that case becomes the linear soil/water partition coefficient described earlier in this chapter, K_D (L/kg).

A theoretical criterion has been given for conditions under which the sorption of pesticides to soil may be treated as if equilibrium prevails (Rao and Jessup 1983). The condition is related to the dimensionless Damkohler number, Da, for sorption and convection, a dimensionless parameter given by:

$$Da = \frac{k_s L}{\lceil q / \theta \rceil} \qquad (4\text{-}9)$$

where L (m) is a characteristic length and k_s (1/day) is the first-order rate constant for nonequilibrium sorption to the unitless fraction, F, of Class 2 sorption sites in soil. In a laboratory column study, L is the length of the soil column. Its selection in field studies is not as clear-cut, but the dispersivity defined in Eq. (4-11) below would be a reasonable choice. Rao and Jessup (1983) stated that when Da is greater than about 5, equilibrium in soil sorption may be assumed. The denominator of Eq. (4-9) is simply the magnitude of the interstitial water velocity through soil.

The mass flux of pesticide through the soil, J (kg m/L day) is described as:

$$J = qC - \theta(D_{dis} + D_{dif}) \frac{dC}{dz} \qquad (4\text{-}10)$$

in which D_{dis} (m^2/day) and D_{dif} (m^2/day) are the coefficients of dispersion and diffusion, respectively. Of these two coefficients, the dispersion coefficient is generally much greater in magnitude (Sadeghi, Kissel, and Cabrera 1989). The dispersion coefficient represents spreading of the pesticide concentration due to the spatial nonuniformities of water movement through the soil profile. It is often assumed proportional to the interstitial velocity of water movement through the soil profile:

$$D_{dis} = L_{dis} |q/\theta| \qquad (4\text{-}11)$$

in which L_{dis} (m) is the characteristic dispersion length or dispersivity.

In the hydrology of pesticide movement through soil, dispersion refers to the spreading of a pesticide plume as it travels through the medium. This spreading is a direct result of the spatial nonuniformities of velocity observed by the various flow paths of individual streams of water through the soil. The theoretical basis for the description of dispersion is a subject of current dispute. In order to understand the issues involved in this dispute, it is helpful to reflect on the original problem leading to the definition of the term dispersion.

G. I. Taylor, a British scientist of the mid-20th century, sought a mathematical description of a phenomenon that he had observed in the laboratory (1953). He had spent some time observing dye movement in fluids flowing at constant velocity and in a laminar (nonturbulent) manner through cylindrical, transparent tubes. His experiment was to introduce pulses of dye as a thin plane across the entire tube and observe the deformation of that sheet and its eventual decomposition into a Gaussian-shaped plume that slowly spread lengthwise while traveling down the tube at the mean velocity of water flow.

The initial deformation of the dye plane is of course caused by the fact that water in the center of the tube is traveling much faster than water at other radial positions. At the very edge of the pipe, the velocity actually falls to zero. The parabolic nature of this velocity profile was well known to Taylor. This movement of the dye, known as convection or advection because it is caused by the motion of water, is the dominant process, but not the only one effecting spreading of the dye. As the initial planar sheet is deformed into a parabolic sheet, very large concentration gradients develop, and the natural tendency of all molecules, as a result of Brownian motion, is to diffuse from areas of high concentration down to areas of low concentration. Thus, the mathematical description of this phenomenon requires two components:

1. An equation describing the velocity profile across the tube; and
2. An equation describing the lateral molecular diffusion of material in response to any concentration gradients that develop.

Taylor was able to show that molecular diffusion in the transverse direction became relatively unimportant at distances not far from the point where the dye was introduced.

Taylor's elegant and experimentally verified mathematical solution to this problem was approximate only in the sense that it applied exclusively for the portion of the pipe removed a distance, L (m), from the point where the dye was introduced:

$$L = \frac{vr^2}{D_o} \tag{4-12}$$

in which v (m/d) is the mean velocity in the pipe, r (m) is the radius of the pipe, and D_o (m2/day) is the molecular diffusion coefficient of the dye in the fluid. This restriction was imposed on the solution because Taylor had explicitly ignored concentration gradients and therefore diffusion in the direction of flow, a simplification that is simply not valid until a certain distance has been covered by the flow in the pipe. The exact analysis at intermediate times has been given (Gill and Sankarasubramanian 1970), in which it was shown that rapid increases in the dispersion coefficient occur until $L > 0.2vr^2/D_o$.

The controversy facing those challenged with describing pesticide dispersion in soil derives from the fact that Taylor's dispersion analysis was adopted to describe the process even though the physics of the situation are very different. Researchers have recently used sophisticated three-dimensional models of pesticide transport in unsaturated heterogeneous soils to show that the dispersion coefficient in realistic field conditions would not be expected to reach the asymptotic value (Liu, Loague, and Feng 1991). Besides not having reached an asymptotic value, there is considerable variability in dispersion parameters fitted to field data for the same soil but under slightly different environmental conditions. In a study using surface-applied bromide in field plots of Oklahoma (Smith, Ahuja, and Ross 1984), values of D_{dis} ranging from 0.0004 to 0.0045 m²/day were fitted to the field data observed in the same plots but in different years.

One possible solution to this theoretical problem is to acknowledge the fact that asymptotic behavior is not generally obtained under real field conditions and to allow L_{dis} to increase linearly with the mean distance traveled (Dieluin, Matherton, and de Marsily 1981). The rate of increase of L_{dis} with mean distance traveled has been related to the Capillary number, N_{ca}, a dimensionless parameter defined as:

$$N_{ca} = \frac{|q/\theta|\eta}{\gamma} \tag{4-13}$$

in which η (kg/m day) is the viscosity of water and γ (kg/day²) is its surface tension. The physical interpretation of the Capillary number is that it expresses a ratio of viscous to surface tension forces, which are the two strongest forces affecting water movement in the unsaturated zone of soil.

A correlation between the characteristic dispersion length, Capillary number, and mean travel distance has been proposed (Gustafson 1988a) as:

$$L_{dis} = (3.8 \times 10^5)L_{mtd} (N_{ca})^{0.6} \tag{4-14}$$

in which L_{mtd} (m) is the mean travel distance covered by the pesticide during the course of the experiment.

The diffusion coefficient is usually negligible in magnitude when compared with the dispersion coefficient, except under very dry conditions of little water movement or with extremely volatile pesticides. It is related to the diffusion coefficient of the pesticide in water, D_o (m^2/day) as follows:

$$D_{dif} = \lambda D_o \qquad (4\text{-}15)$$

in which λ is the dimensionless tortuosity factor.

The continuity equation for the pesticide in soil is given as:

$$\frac{dC^*}{dt} = -\frac{dJ}{dz} - R_c \qquad (4\text{-}16)$$

in which R_c (kg/L day) is the net dissipation rate for the pesticide.

It is often assumed that linear, first-order reaction kinetics hold for the net dissipation rate of the pesticide, hence:

$$R_c = k \, C^* \qquad (4\text{-}17)$$

in which k is the first-order rate constant for the dissipation reaction.

As discussed earlier in this chapter, the dissipation of most pesticides in soil is not well-described by linear first-order kinetics. This has been explained to be a natural result of spatial variability in the first-order rate constant (Gustafson and Holden 1990). In this solution to the dissipation rate equation, a Gamma distribution was assumed to describe the spatial variability of the first-order constant. The gamma (Γ) is a common distribution for nonnegative random variables, whose pdf is given by:

$$f(k) = \frac{k^{a-1}e^{-k/\beta}}{\beta^a \Gamma(a)} \quad a > 0, \beta > 0, k > 0 \qquad (4\text{-}18)$$

in which a (unitless) and β (1/day) are simply the two parameters of the gamma (Γ) distribution describing the variability of the rate constant.

The mean of the gamma distribution is given by the product $a\beta$, and the variance is $a\beta^2$. The solution to the dissipation rate equation assumes the following simple form when the first-order rate constant varies spatially according to Eq. (4-18):

$$C^* = C_o (1 + \beta t)^{-a} \qquad (4\text{-}19)$$

in which C_o is the initial concentration. It should be noted that as β vanishes this equation reduces to the simple linear first-order dissipation model, and the gamma distribution becomes degenerate (that is, all of the rate constants have the same value).

In order to be incorporated into the differential equation (Eq. [4-16]) describing the transport of pesticide through soil, this nonlinear equation must be expressed in a differential form not involving the gamma distribution. Simple differentiation of Eq. (4-19) with respect to time yields:

$$\frac{dC^*}{dt} = -\frac{\alpha\beta C^*}{1 + \beta t} \tag{4-20}$$

which can be set equal to the R_c term in Eq. (4-14). Values of a in this rate equation are often near 1, and β can be set equal to the inverse of the soil half-life or, more properly, DT_{50}. Equations relating DT_{50} to a and β for the general case are given by Gustafson and Holden (1990).

Besides the effect of spatial variability on the rate of dissipation, a number of other factors can be introduced into the pesticide dissipation model. One of the most complete is that proposed by Boesten and Van Der Linden (1991):

$$k = f_T f_\theta f_z k_{ref} \tag{4-21}$$

in which f_T is a unitless factor for the influence of soil temperature, f_θ is a unitless reduction factor for the influence of soil moisture, f_z is a unitless reduction factor for the influence of depth in soil, and k_{ref} (1/day) is the first-order rate constant for dissipation at some convenient reference condition. The reference conditions proposed by Boesten and Van Der Linden were at a temperature of 20°C and at a moisture condition leading to a hydraulic potential of -10 kPa.

The temperature factor, f_T, is generally described using the Arrhenius equation:

$$f_T = exp\left[\frac{E_a(T - T_{ref})}{R\,T\,T_{ref}}\right] \tag{4-22}$$

in which E_a (cal/mol) is the activation energy for the dissipation reaction and R (1.987 cal/mol °K) is the universal gas constant. The temperatures in this equation must be expressed on the absolute scale, °K. Over the rather narrow temperature range that soil typically exhibits (0–30°C or 273–303°K), the factor given by $E_a/(RTT_{ref})$ is nearly a constant. Thus, Eq. (4-22) may be simplified greatly collecting these terms as a constant, κ (1/°C), yielding:

$$f_T = exp(\kappa(T - T_{ref})) \tag{4-23}$$

in which the two temperatures need no longer be given on the absolute scale, since it is only the deviation from the reference condition that is required.

The factor f_θ was given by Boesten and Van Der Linden as:

$$f_\theta = \min[1, (\theta/\theta_{ref})^B] \tag{4-24}$$

in which "min" means "the minimum of," θ_{ref} is the soil moisture content at the assumed reference conditions, and B is a unitless constant. This functional dependence of the dissipation rate on soil moisture content was first proposed by Walker (1976), who found that B was typically in the range of 2 to 3. By using the "min" function in Eq. (4-24), it is assumed that the soil moisture content has no influence on dissipation rate at conditions wetter than the reference conditions, which is generally consistent with the available data. The most drastic effects of moisture are observed when the soil dries out to very low moisture contents, which is the effect caused by the use of B equal to 2 or 3 in Eq. (4-24).

Various functional forms for the effect of depth in soil on dissipation rate, f_z, have been proposed. Jury, Focht, and Farmer proposed (1987) that depth in soil had no effect until a critical depth, z_{crit} (m), was reached, below which there was an exponential damping effect on dissipation rate until a residual dissipation rate was attained at an even deeper soil depth, z_r (m):

$$f_z = \exp[-\zeta(z - z_{crit})] \quad for \ z_r > z > z_{crit} \tag{4-25}$$

in which ζ (1/m) is the damping factor for dissipation. Values for these parameters assumed by Jury were $z_{crit} = 1$ meter, $\zeta = 3$ meters, and $z_r = 3$ meters.

Boesten and Van Der Linden gave a piecewise linear function for $f_z(z)$, which began at unity down to approximately 0.25 meters and then angled down to 0 in three separate segments at a depth of 1 meter.

Other factors thought to influence the rate of dissipation include soil pH, organic matter, and biomass. None of the published models have explicitly included effects such as these on the rate of dissipation.

Collecting all of the above information into one single equation is not possible; however, the common starting pointing for most pesticide transport models is some form of the convective-dispersion equation (often abbreviated CDE), which involves the substitution of Eq. (4-10) and a dissipation model such as Eq. (4-20) into Eq. (4-16), yielding:

$$\frac{dC^*}{dt} = -q\frac{dC}{dz} - \theta(D_{dis} + D_{dif})\frac{d^2C}{dz^2} - \frac{\alpha\beta C^*}{1 + \beta t} \tag{4-26}$$

in which an appropriate model for the dispersion coefficient should be substituted and the diffusion coefficient can often be ignored and set equal to zero.

The above development is concerned only with the rate of dissipation of the parent pesticide. In many cases, such as with aldicarb, it is the concentration of daughter products that must be accounted for. A kinetic model of aldicarb dissipation in soil has been described (Ou et al. 1988). A schematic of this reaction scheme and definitions of the individual terms are shown in Figure 4-4.

Assuming first-order kinetics for each reaction, the differential equations describing the concentrations of each of the three carbamates are as follows:

$$\frac{dA}{dt} = -[k_1A + k_3A] = -k_{13}A \tag{4-27}$$

$$\frac{dB}{dt} = k_1A - [k_2B + k_4B] = k_1A - k_{24}B \tag{4-28}$$

$$\frac{dC}{dt} = k_2B - k_5C \tag{4-29}$$

in which the following terms are used: $k_{13} = [k_1 + k_3]$ and $k_{24} = [k_2 + k_4]$.

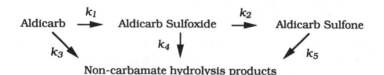

A = concentration of aldicarb (kg/ha soil)

B = concentration of aldicarb sulfoxide (aldicarb equivalent kg/ha soil)

C = concentration of aldicarb sulfone (aldicarb equivalent kg/ha soil)

k_1 = oxidation rate constant (1/day) for aldicarb

k_2 = oxidation rate constant (1/day) for aldicarb sulfoxide

k_3 = hydrolysis rate constant (1/day) for aldicarb

k_4 = hydrolysis rate constant (1/day) for aldicarb sulfoxide

k_5 = hydrolysis rate constant (1/day) for aldicarb sulfone

FIGURE 4-4.   Kinetic model for aldicarb and its carbamate degradates.

Analytical solutions to Eqs. (4-27) through (4-29), subject to the initial condition that $A = A_o$, $B = 0$, and $C=0$, are given by the following:

$$A = A_oX_1 \tag{4-30}$$

$$B = \left(\frac{k_1A_o}{b_1}\right)(X_1 - X_2) \tag{4-31}$$

$$C = \left(\frac{k_1k_2A_o}{b_1}\right)\left(\frac{b_1X_3}{b_2b_3} + \frac{X_1}{b_3} - \frac{X_2}{b_2}\right) \tag{4-32}$$

in which the following terms are used: $b_1 = [k_{24} - k_{13}]$, $X_1 = exp[-k_{13}t]$, $b_2 = [k_5 - k_{24}]$, $X_2 = exp[-k_{24}t]$, $b_3 = [k_5 - k_{13}]$, and $X_3 = exp[-k_5t]$.

The equations and mathematical theory presented here represent only a selection of the various mathematical approaches that have been given in the literature. As discussed earlier in the chapter, van Genuchten and Wagenet (1989) have given analytical solutions to the flexible two-site/two-region model, incorporating more recent thinking concerning pesticide sorption and transport through soil. To date, however, such increases in model complexity have not been incorporated into widely available models describing pesticide contamination of drinking water, the topic of next concern.

MODELS DESCRIBING PESTICIDE CONTAMINATION OF DRINKING WATER

Not long after the reports emerged of pesticide residues being found in drinking water, a host of computer models purporting to describe the process began to appear. These programs vary widely in complexity from simple compartmental models to full-fledged simulations of entire watersheds involving scores of coupled differential equations. One common feature of all but a select few of these models is that they have been used very little after having been developed, and in these few cases the user has almost always been the author of the model.

Some of the models to be identified here are simply numerical inequalities based on the physical properties, while others are full-fledged simulations. An intermediate modeling approach is the use of generalized probability models (Di Toro 1982; Mills and Leonard 1984). In this technique, the water quality response of a stream receiving intermittent storm runoff is addressed within a probabilistic framework. Lognormally distributed runoff and stream flows and concentrations are assumed. These may be used to assess whether it is necessary to embark upon more detailed time variant or Monte Carlo simulations.

The most widely used models are discussed below. Further details about obtaining and operating certain of these programs are given in Appendix 2.

PRZM/RUSTIC

Written by Robert Carsel and others (1984) at the EPA's research laboratory in Athens, Georgia, the Pesticide Root Zone Model (PRZM) has been widely used by a number of researchers interested in predicting pesticide contamination of drinking water. Reported uses of PRZM have included a number of purposes, ranging from the development of geographic screening indices to the prediction of surface water runoff potential, and even the site-specific leaching simulations for which it was originally intended. Although it has been roundly castigated by various purists in the academic world for its multiple and often unjustifiable simplifications, PRZM stands alone among comprehensive site simulation models in terms of its ease of use, completeness of documentation, and lack of "bugs" or computer errors preventing the program from running. Its utility for anyone hoping to study the potential for drinking water contamination by pesticides is undeniable.

More recently (Dean et al. 1989), the PRZM model has been upgraded and incorporated within a much more complex program called RUSTIC, which stands for Risk of Unsaturated/Saturated Transport and Transformation of Chemical Concentrations. The RUSTIC model was originally composed of three component models: PRZM, VADOFT, and SAFTMOD; however, support for SAFTMOD has recently been dropped. PRZM and VADOFT are one-dimensional models representing the vertical transport of pesticides through the root and vadose zones, respectively. SAFTMOD was a two-dimensional saturated zone transport model for examining the flow of pesticides once they reached the groundwater. The RUSTIC code was designed to provide state-of-the-art deterministic simulation of the fate and transport of pesticides, applied for agricultural practices, in the root, vadose, and saturated zone. The model is capable of simulating multiple pesticides or parent/daughter relationships. The model is also capable of estimating probabilities of concentrations or fluxes in these various media for the purpose of performing exposure assessments.

In a careful comparison of observed chemical movement in laboratory soil columns with the predictions of three computer models (Melancon, Pollard, and Hern 1986), PRZM gave substantially more accurate predictions than two other programs, SESOIL and PESTANS, although it did not give good agreement unless site-specific values for pesticide persistence and mobility were utilized. In a comparison of PRZM model predictions with observed field behavior (Carsel, Nixon, and Ballantine 1986), it was found that the computer program effectively simulates the important processes operating on the pesticide metalaxyl in the unsaturated zone. However, in order to improve model agreement with observed data, it was found necessary to lower the first-order degradation rate partway through the study, because of an apparent slowdown in degradation rate.

Further details about the inner workings of PRZM/RUSTIC and instructions for obtaining a copy of the program are given in Appendix 2.

EXAMS

EXAMS (Exposure Analysis and Modeling System) is an interactive modeling system that allows a user to specify and store the properties of chemicals and ecosystems, modify either via simple commands, and conduct rapid evaluations and error analyses of the probable aquatic fate of synthetic organic chemicals (Burns 1989). EXAMS combines the loadings, transport, and transformations of a chemical into a set of differential equations, using the law of conservation of mass as an accounting principle. It accounts for all the chemical mass entering and leaving a system as the algebraic sum of external loadings, transport processes that export the compound from the system, and transformation processes within the system that convert the chemical to daughter products. The program produces output tables and simple graphics describing:

1. Chemical exposure, the expected environmental concentrations (EECs) resulting from a specified pattern of chemical loadings;
2. Fate, the distribution of the chemical in the system and the fraction of the loadings consumed by each transport and transformation process; and
3. Persistence, the time required for purification of the system (via export/transformation processes) should the chemical loadings cease.

EXAMS includes process models of the physical, chemical, and biological phenomena governing the transport and fate of compounds. This set of unit process equations used to compute the kinetics of chemicals is the central core of EXAMS. Each equation expresses in mathematical form the interactions between the chemistry of a compound and the environmental forces that shape its behavior in aquatic systems. This "second-order" or "system-independent" approach lets one study the fundamental chemistry of compounds in the laboratory and then, based on independent studies of the levels of driving forces in aquatic systems, evaluate the probable behavior of the compound in systems that have never been exposed to it. Most of the process equations are based on standard theoretical constructs or accepted empirical relationships. Ionization of organic acids and bases and sorption of the compound with sediments and biota are treated as thermodynamic properties or (local) equilibria that constrain the operation of the kinetic processes.

EXAMS allows for the simultaneous treatment of up to 28 molecular species of a chemical, including the parent uncharged molecule, and singly, doubly, or triply charged cations and anions, each of which can occur in a dissolved, DOC-complexed, sediment-sorbed, or biosorbed form. The user can specify reaction pathways for the production of transformation products of concern, whose further fate and transport can then be simultaneously simulated by EXAMS.

EXAMS computes the kinetics of transformations attributable to direct photolysis, hydrolysis, biolysis, and oxidation reactions. The input chemical data for hydrolytic, biolytic, and oxidative reactions can be entered either as single valued, second-order rate constants or as pairs of values defining the rate constant as an Arrhenius function of the environmental temperature specified for each segment. EXAMS has been designed to accept standard water-quality parameters and system characteristics that are commonly measured by limnologists throughout the world as well as chemical datasets conventionally measured or required by EPA regulatory procedures.

In a field study comparing EXAMS predictions with the observed field behavior of the insecticide endothall (Reinert, Rocchio, and Rodgers 1987), reasonable agreement was obtained. However, the workers found that only one process variable, the biotransformation rate constant, was really being tested in the comparison. The volatilization routines utilized by EXAMS were validated in model pond systems for several pesticides (Sanders and Seiber 1984).

As with the previous model, further details about the inner workings of EXAMS and instructions for obtaining a copy of the program are given in Appendix 2.

SWRRB

SWRRB (Simulator for Water Resources in Rural Basins) was developed for simulating hydrologic and related processes in rural basins. The objective in model development was to predict the effect of management decisions on water, sediment, nutrient, and pesticide yields with reasonable accuracy for rural basins throughout the United States. To satisfy this objective, the model had to be:

1. Physically based and use readily available inputs;
2. Capable of computing the effects of management changes on outputs;
3. Computationally efficient, to allow simulation on a variety of management strategies without excessive cost;

4. Capable of simulating long periods for use in frequency analysis; and
5. Capable of operating on a subdivided basis (or subbasin scale).

The last feature is critical when attempting to model complex sites.

The major processes included within SWRRB are surface runoff, percolation, return flow, evapotranspiration, transmission losses, pond and reservoir storage, sedimentation, and crop growth. As in PRZM, the SCS curve number technique is used to predict surface runoff. The Modified Universal Soil Loss Equation (MUSLE) is used to model erosion losses. The main difference between SWRRB and PRZM regarding pesticide transport is the much simplified technique (only a single surface layer is considered by SWRRB) for tracking the vertical movement of pesticide, which is due to the fact that SWRRB was never intended to model leaching.

The model has, at various times in the past, been the model of choice by the EPA when estimating loads of pesticide to surface water. At this time, there has been a somewhat reduced use of the program for this purpose, but development work continues in an attempt to make SWRRB more "user-friendly." This may result in more use of the program in the future.

CREAMS/GLEAMS

CREAMS (Chemicals, Runoff, and Erosion, from Agricultural Management Systems) is a physically based, daily simulation model written by USDA ARS scientists that estimates runoff, erosion/sediment transport, plant nutrient, and pesticide yield from field-sized areas. CREAMS was the predecessor of GLEAMS (Ground Water Loadings Estimation from Agricultural Management Systems), which represented an extension of the models capabilities to account for leaching quantities of pesticide in addition to the runoff quantities already predicted by CREAMS.

The CREAMS/GLEAMS model is probably the second-most popular pesticide transport model after PRZM. Several field studies have shown the "reasonableness" of its predictions. Accurate prediction of toxaphene field runoff by the CREAMS model has been reported (Lorber and Mulkey 1982). Similarly, good agreement with field results was reported in a comparison of GLEAMS predictions with field data collected near Tifton, Georgia (Leonard et al. 1988). A comparison of groundwater loadings predicted by GLEAMS with observed concentrations in shallow wells in the Atlantic coastal plain showed that the model predicted concentrations 3 to 7 times higher than observed (Shirmohammadi et al. 1989). Despite this discrepancy, the authors felt that the results support the use of GLEAMS for comparing managerial effects on pesticide movement to groundwater if appropriate limitations are recognized.

The hydrologic component of the program consists of two options. When only daily rainfall data are available to the user, the SCS curve number model is used to estimate surface runoff. If hourly or break-point rainfall are available, an infiltration-based model is used to simulate runoff. Both methods estimate percolation through the root zone of the soil. The erosion component maintains elements of the Universal Soil Loss Equation (USLE), as in PRZM, but includes sediment transport capacity for overland flow. A channel erosion/deposition feature of the model permits consideration of concentrated flow within a field. Impoundments are treated in the erosion component also. The plant nutrient sub-model of CREAMS/GLEAMS has a nitrogen component that calculates mineralization, nitrification, and denitrification of the root zone. Both the nitrogen and phosphorous parts of the nutrient component use enrichment ratios to estimate that portion of the two nutrients transported with sediment. The pesticide component considers foliar interception, degradation, and washoff, as wells as sorption, desorption, and degradation in the soil. This method, like the nutrient model, uses enrichment ratios and partition coefficients to calculate the separate sediment and aqueous phases of pesticide loss.

GLEAMS has been used to develop a simple screening model prediction of the potential for ground and surface water contamination by pesticides (Leonard and Knisel 1988). The form of the correlation is simple: $A_g + (B_g DT_{50}/K_{OC})$. The parameters A_g and B_g in this equation were related to soil type and pesticide water solubility.

SURFACE

The surface water monitoring conducted by Monsanto in the mid-1980s prompted the development, calibration, and validation of the SURFACE model for predicting the concentrations of pesticides in the rivers of intensively farmed regions (Gustafson 1990). In water systems sampled by Monsanto during 1985 and 1986, SURFACE predictions of annualized mean concentrations for alachlor, atrazine, cyanazine, and metolachlor were within 0.09 ppb half of the time.

SURFACE-utilized PRZM simulations to predict edge-of-field runoff loadings of pesticides. These edge-of-field loadings were then routed through the watershed in order to model concentrations at a water withdrawal point downstream of the treated fields. Watersheds ranging in size from only a few square miles to thousands of square miles were each discretized into square mile parcels to form a distributed model of the river basin. In a comparison of lumped and distributed models for chemical transport by chemical runoff (Emmerich, Woolhiser, and Shirley 1989), it was found that total chemical transport was virtually unaffected by model form. However, significant differences were found in arrival times and peak concentrations. The lumped models, which ignored spatial variations of chemical concentration in overland flow and in the surface mixing zone, predicted lower peak concentrations and arrival times that were too short.

Calibration of SURFACE was performed by forcing agreement between its predictions and concentrations of alachlor, atrazine, cyanazine, and metolachlor found during the 1985 surface water survey. Preliminary validation was performed by comparing predictions of the model with 1986 monitoring results for the same four chemicals. The model was most accurate under moderate to low rainfall conditions. Highest concentrations are predicted and observed for intensively farmed watersheds of low hydraulic permeability in which reservoirs of long residence time are utilized as the source of drinking water.

One of the most important factors in the SURFACE model is the fraction of edge-of-field pesticide runoff that actually reached the major streams of the watershed, which was regarded as an adjustable parameter. Previous comparison of other model predictions with surface water data had suggested that only 10 percent of the edge-of-field runoff reached the streams. This value (10 percent) was used for all four chemicals in all watersheds and appeared to give reasonable agreement with the 1985 data. No further adjustment to the parameter was attempted. As reported by Gustafson (1990), it appears that this parameter should be adjusted upward under very wet conditions, during which slightly more than 10 percent of the edge-of-field runoff is apparently able to reach the streams of the watershed.

The SURFACE model gives only an approximate representation of the entire weekly time series, but it gives reasonably reliable predictions of the timing and size of the peak value. More accurate predictions of the maximum would probably require more detailed information on application dates within the watershed.

HSPF/STREAM

Another model for predicting concentrations of pesticides in surface water is the Hydrologic Simulation Program–FORTRAN (HSPF) program that was developed in the 1970s and 1980s under contract for the EPA (Donigian and Mulkey 1987). This highly complex model includes over 50,000 lines of scantily commented FORTRAN code and requires no fewer than 300 input parameters. It requires hourly weather data on a regional scale, data that are almost never available. In many ways, HSPF typifies what not to do when developing a computer model of pesticide transport (or of any other process, for that matter).

Not only complex, the model was never compared by its authors with any complete set of field data to determine whether it gave a reasonable prediction of reality. In fact, the only time the model was used extensively was when its developers were contracted, again by the EPA, to predict what concentrations of alachlor would likely occur in the watersheds where the herbicide is used in the midwestern United States. In this use of their own model, the authors predicted average annual concentrations of alachlor in midwestern rivers of 15 μg/L. Monitoring data, as pointed out in Chapter 2, showed this prediction to be at least one order of magnitude high.

The only other use of HSPF reported was its use to develop STREAM, a series of nomograms for rapidly assessing the likely concentrations of pesticides in surface water (Donigian and Mulkey 1987). The nomograms are based on knowledge of the application rate, market share, DT_{50}, and K_{OC} of the pesticide. If the accuracy of HSPF could be accepted, then one could possibly accept the alleged "order-of-magnitude" accuracy of STREAM. In fact, as discussed above, detailed comparison of actual surface water quality monitoring data with the predictions of HSPF demonstrated that it was predicting concentrations about an order-of-magnitude high. Where this leaves the predictions of STREAM is uncertain, but one comparison revealed the prediction to be erroneously high (Gustafson 1988b).

GUS

GUS (Groundwater Ubiquity Score) is a screening technique for rapidly assessing whether a given pesticide is likely to be a contaminant of drinking water (Gustafson 1989). It was developed from a group of pesticides that had been classified by the state of California's Department of Food and Agriculture (CDFA) as either contaminants or non-contaminants of groundwater. The CDFA had also collected physical properties for the same set of pesticides. These data were used to combine persistence and mobility measures into an index that effectively separated the two groups of chemicals:

$$GUS = (log_{10}DT50) * (4 - log_{10} K_{OC}) \qquad (4\text{-}33)$$

for which values of GUS > 2.8 indicate a high probability that the pesticide will be a contaminant, whereas GUS < 1.8 indicates a very low probability of it being a contaminant.

The GUS index has been used to develop an overall economic model of the net societal impacts of pesticide use on water quality, farm produce cost, and growers' production costs (Hoag 1990; Hoag and Hornsby 1991). Other uses have included the development of a screening method in Canada (see

Chapter 5) and in an analysis of the National Pesticide Survey results by the EPA (see Chapter 2).

Another application involving the use of the GUS index is one by the USDA Soil Conservation Service (SCS) which used it as part of their screening procedure to determine whether the use of a particular pesticide on a particular soil would be likely to result in load to either ground or surface water. Once developed, the SCS procedure is to be used by USDA Extension Service personnel at the county level to advise farmers on whether it is prudent to use particular pesticides on their fields. The SCS screening procedure was first introduced in 1988 and was subsequently modified to improve the description of chemical runoff (Goss and Wauchope 1990). A computerized version of the system, NPURG, was developed for use in Massachusetts by Jeff Jenkins, now at Oregon State University. The physical properties of pesticides needed by both the original SCS methodology and NPURG come from the USDA ARS/SCS pesticide properties data base (Wauchope et al. 1992). The information in this data base was obtained both from the literature and by direct submission of proprietary results of the pesticide manufacturers to the USDA.

Jury Screening Model

In 1987, William Jury and coauthors published a screening model for evaluating the groundwater contamination potential of a pesticide (Jury, Focht, and Farmer 1987). While similar in some ways to the GUS screening index, the model was developed in an entirely different manner. The analytical solution to a simplified leaching model was derived and subsequently simplified to a mathematical inequality of the following form:

$$K_{OC} > a\,(DT50) - b \tag{4-34}$$

where a (L/kg day) and b (L/kg) are parameters related to the specific scenario modeled. For a low pollution potential region, the following pair of parameters was given: $a = 0.00104$ L/kg day, $b = 14$ L/kg. For a high pollution potential region, the parameters were specified as: $a = 0.0159$ L/kg day, $b = 27$ L/kg. As DT_{50} increases, Eq. (4-34) states that the maximum K_{OC} is needed to avoid leaching problem increases.

The approach taken in the Jury screening model was further developed by Boesten and van der Linden (1991). Rather than utilizing an over-simplified leaching model in which dispersion was completely ignored, they utilized a full-fledged numerical solution to the convective-dispersion equation. They reported similar results to Jury and his coworkers and did not specifically demonstrate any practical advantages of their technique. Another extension to this approach that has been reported is to incorporate a health-based maximum allowable concentration such as MCL or HAL as a measure of risk (Steenhuis and Taylor 1987).

DRASTIC

While not a simulation model or even a pesticide-specific screening technique, the DRASTIC groundwater vulnerability assessment scheme has been used by several researchers in evaluating the potential risks of pesticides to groundwater quality. The DRASTIC initials stand for: D—depth to groundwater, R—recharge rate, A—aquifer material, S—soil type, T—topography, I—impact of the vadose zone, and C—conductivity, as in hydraulic conductivity of the aquifer material (Aller et al. 1987).

Although it has now been shown (see Chapter 2) to have performed quite poorly in the EPA's National Pesticide Survey and in Monsanto's National Alachlor Well Water Study, DRASTIC continues to be used. For instance, it was recently used to identify sampling sites in Utah (Ehteshami et al. 1991). In another recent effort, National Ground Water Association members were surveyed in order to develop a data base of 400 field sites classified according to the DRASTIC system (Newell, Hopkins, and Bedient 1990). The results indicate that the EPA's assumed distributions of seepage velocity and hydraulic conductivity are sound. The data base is available as a spreadsheet computer file from the American Petroleum Institute.

In another assessment of the DRASTIC technique, PRZM-based predictions of the total quantity of pesticide leachate for a region were compared with DRASTIC scores (Banton and Villeneuve 1989). No significant correlation was found, and the authors concluded that a number of factors not considered by DRASTIC are more important than many of the factors making up the DRASTIC score.

Multimedia Models

Multimedia models involving the fate of pesticides in air, soil, and water have been described for several years by Donald Mackay at the University of Toronto (Mackay 1979; Mackay and Peterson 1981, 1982, 1984, 1991; Southwood, Harris, and Mackay 1989). Such techniques have found more extensive use in studying the global transport of pesticides, rather than as a tool to understand the potential for drinking water contamination in a particular region (Eschenroeder 1983).

NOTATION USED IN THE EQUATIONS OF CHAPTER 4

Roman Symbols

a = unitless factor in the Jury screening model

b = factor in the Jury screening model

B = unitless factor giving soil moisture dependence of pesticide dissipation in soil

C = concentration of pesticide in soil water (kg/L)

C_o = initial concentration of pesticide in soil (kg/L)

$C*$ = total concentration of pesticide in soil (kg/L)

D_{dif} = diffusion coefficient of pesticide in soil (m²/day)

D_{dis} = dispersion coefficient of pesticide in soil (m²/day)

D_L = dispersion coefficient defined by the Taylor dispersion analysis (m²/day)

D_o = diffusion coefficient of pesticide in water (m²/day)

Da = unitless Damkohler number giving the relative importance of sorption kinetics in the transport of pesticides through soil

DT_{50} = time required for the first 50 percent of the applied pesticide to dissipate (day)

DT_{90} = time required for the first 90 percent of the applied pesticide to dissipate (day)

E = rate of evaporation of water at soil surface (m/day)

E_a = activation energy for the dissipation of pesticide in soil (cal/mol)

E_p = potential rate of evaporation of water (m/day)

f_{oc} = unitless fraction organic carbon content of soil

f_T = unitless factor for the influence of temperature on the pesticide dissipation rate

f_z = unitless reduction factor for the influence of depth in soil on the pesticide dissipation rate

f_θ = unitless reduction factor for the influence of soil moisture on the pesticide dissipation rate

F = unitless fraction of equilibrium-type sorptive domains in soil

GUS = unitless Groundwater Ubiquity Score

J = mass flux of pesticide through soil (kg m/L day)

k = first-order rate constant for pesticide dissipation in soil (1/day)

k_{ref} = first-order rate constant for pesticide dissipation in soil at reference conditions (1/day)

k_s = first-order rate constant for pesticide sorption to the rate-limited soil sorptive domains (1/day)

K_D = soil-water partition coefficient (L/kg)

K_f = Freundlich soil sorption coefficient ($L^{1/n}/kg^{-1/n}$)

K_h = hydraulic conductivity of soil (L/kg day)

K_{OC} = soilwater partition coefficient based on soil organic carbon content (L/kg)

L = distance in pipe required for Taylor's dispersion analysis to apply (m)

L_{dis} = dispersivity or characteristic dispersion length (m)

L_{mtd} = mean distance traveled by pesticide in soil (m)

n = unitless Freundlich soil sorption exponent

N = volumetric rate of precipitation (m/day)

N_{ca} = unitless Capillary number expressing relative importance of viscous and surface tension forces in unsaturated soil

P_h = hydraulic pressure (kg/m day)

q = volumetric rate of water movement (m/day)

r = radius of pipe in the Taylor dispersion analysis (m)

R = universal gas constant, 1.987 cal/mol K

R_c = net dissipation rate of pesticide in soil (kg/L day)

S = rate of water uptake by crop (1/day)

S_p = potential rate of uptake of water by crop (1/day)

t = time (days)

T = temperature (K)

T_{ref} = reference temperature (K)

v = velocity of water in pipe in the Taylor dispersion analysis (m/day)

X = quantity of pesticide sorbed to soil (kg/kg)

z = depth in the soil profile (m)

z_{crit} = critical depth in soil at which pesticide dissipation rate begins to decline (m)

z_r = crop rooting depth (m)

z_{res} = depth in soil at which pesticide dissipation rate attains residual value (m)

z_{sat} = water table depth (m)

Greek Symbols

α = unitless shape factor of the Gamma distribution for the dissipation rate constant

β = scaling factor of the Gamma distribution for the dissipation rate constant (1/day)

ε = unitless efficiency of crop uptake of water

γ = surface tension of water (kg/day^2)

$\Gamma(a)$ = Gamma function of a,

η = viscosity of water (kg/m day)

κ = factor expressing temperature dependence of pesticide dissipation rate in soil (1/C)
λ = unitless tortuosity factor of diffusion paths in soil
θ = unitless volume fraction of water in soil
θ_{ref} = unitless volume fraction of water in soil at reference conditions
θ_{sat} = unitless saturated volume fraction of water in soil
ρ_s = dry bulk density of soil (kg/L)
ζ = factor expressing depth dependence of pesticide dissipation rate in soil (1/m)

References

Alexander, M., and K.M. Scow. 1989. Kinetics of biodegradation in soil. In *Reactions and Movement of Organic Chemicals in Soils,* ed. B.L. Sawhney, K. Brown, p. 243. Madison, Wis.: Soil Science Society of America and American Society of Agronomy.

Aller, L., T. Bennet, J. Lehr, R. Petty, and G. Hackett. 1987. *DRASTIC, A Standardized System for Evaluating Ground Water Pollution Potential Using Hydrogeologic Settings,* EPA-600/2-87-035.

Alvarez-Cohen, L., and P.L. McCarty. 1991a. A cometabolic biotransformation model for halogenated aliphatic compounds exhibiting product toxicity. *Environ. Sci. Technol.* 25:1381–87.

Alvarez-Cohen, L., and P.L. McCarty. 1991b. Two-stage dispersed-growth treatment of halogenated aliphatic compounds by cometabolism. *Environ. Sci. Technol.* 25:1387–93.

Amoozefar-Fard, A., D.R. Nielsen, and A.W. Warrick. 1982. Soil solute concentration distributions for spatially varying pore water velocities and apparent diffusion coefficients, *Soil Sci. Soc. Am. J.* 46:3–8.

Armstrong, A.Q., R.E. Hodson, H.M. Hwang, and D.L. Lewis. 1991. Environmental factors affecting toluene degradation in ground water at a hazardous waste site. *Environ. Toxic. & Chem.* 10:147–58.

Baker, J.L. 1979. Agricultural areas as nonpoint sources of pollution. In *Environmental Impact of Nonpoint Source Pollution,* ed. M.R. Overcash and J.M. Davidson, pp. 275–310. Ann Arbor: Ann Arbor Science.

Ball, W.P., and P.V. Roberts. 1991. Long-term sorption of halogenated organic chemicals by aquifer material. *Environ. Sci. Technol.* 25:1223–49.

Banton, O. and J.-P. Villeneuve. 1989. Evaluation of groundwater vulnerability to pesticides: a comparison between the pesticide drastic index and the PRZM leaching quantities. *J. Contam. Hyd.* 4:285–96.

Barrio-Lage, G.A., F.Z. Parsons, and P.A. Lorenzo. 1988. Inhibition and stimulation of trichloroethylene biodegradation in microaerophilic microcosms. *Environ. Toxic. Chem.* 7:889–95.

Bilkert, J.N., and P.S.C. Rao. 1985. Sorption and leaching of three nonfumigant nematicides in soils. *J. Environ. Sci. Health* B20:1–26.

Blair, A.M., and T.D. Martin. 1988. A review of the activity, fate and mode of action of sulfonylurea herbicides. *Pestic. Sci.* 22:195–219.

Boesten, J.J.T.I., and A.M.A. Van Der Linden. 1991. Modeling the influence of sorption and transformation on pesticide leaching and persistence. *J. Environ. Qual.* 20:425–91.

Boesten, J.J.T.I. 1991. Sensitivity analysis of a mathematical model for pesticide leaching to groundwater. *Pestic. Sci.* 31:375–88.

Boesten, J.J.T.I., L.J.T. Van Der Pas, and J.H. Smelt. 1989. Field test of a mathematical model for non-equilibrium transport of pesticides in soil. *Pestic. Sci.* 25:187–203.

Braverman, M.P., T.L. Lavy, and C.J. Barnes. 1986. The degradation and bioactivity of metolachlor in the soil. *Weed Science* 34:479–84.

Bromilow, R.H., and M. Leistra. 1980. Measured and simulated behavior of aldicarb and its oxidation products in fallow soils. *Pestic. Sci.* 11:389–95.

Bromilow, R.H., G.G. Briggs, M.R. Williams, J.H. Smelt, L.G.M.T. Tuinstra, and W.A. Traag. 1986. The role of ferrous ions in the rapid degradation of oxamyl, methomyl and aldicarb in anaerobic soils. *Pestic. Sci.* 17:535–47.

Bromilow, R.H., R.J. Baker, M.A.H. Freeman, and K. Gorog. 1980. The degradation of aldicarb and oxamyl in soil. *Pestic. Sci.* 11:371–78.

Brown, H.M. 1990. Mode of Action, Crop Selectivity, and Soil Relations of the Sulfonylurea Herbicides. *Pestic. Sci.* 29:263–81.

Burns, L.A. 1989. *Exposure Analysis Modeling System, User's Guide for EXAMS Version 2.92.* Athens, Ga U.S. EPA.

Carsel, R.F., L.A. Mulkey, M.N. Lorber, and L.B. Baskin. 1984. The Pesticide Root Zone Model (PRZM): A Procedure for Evaluating Pesticide Leaching Threats to Ground Water. Athens, Ga.: U.S. EPA. 22 pp.

Carsel, R.F., W.B. Nixon, and L.G. Ballantine. 1986. Comparisons of pesticide root zone model predictions with observed concentrations for the tobacco pesticide metalaxyl in unsaturated zone soils. *Environ. Toxic. Chem.* 5:345–53.

Chiou, C.T., D.E. Kile, T.I. Brinton, R.L. Malcolm, J.A. Leenheer, and P.L. MacCarth. 1987. A comparison of water solubility enhancements of organic solutes by aquatic humic materials and commercial humic acids. *Environ. Sci. Technology* 21:1231–34.

Cho, H.J., and P.R. Jaffe. 1990. The volatilization of organic compounds in unsaturated porous media during infiltration. *J. Contam. Hydrol.* 6:387–410.

Dao, T.H., and T.L. Lavy. 1987. A kinetic study of adsorption and degradation of aniline, benzoic acid, phenol, and diuron in soil suspensions. *Soil Science* 143:66–72.

Dean, J.D., and R.F. Carsel. 1988. *A Linked Modeling System for Evaluating Impacts of Agricultural Chemical Use.* Athens, Ga: U.S. EPA. 25 pp.

Dean, J.D., P.S. Huyakorn, A.S. Donigian, K.A. Voos, R.W. Schanz, Y.J. Meeks, and R.F. Carsel. 1989. *Risk of Unsaturated/Saturated Transport and Transformation of Chemical Concentrations (Rustic).* Athens, Ga.: U.S. EPA. EPA/600/3-89/048a and B.

Di Toro, D.M. 1982. *Probability Model of Stream Quality Due to Runoff, Proceedings: Impacts of Urban Runoff on Receiving Waters.* 2 Jun 1982. Philadelphia: American Geophysical Union. pp. 607–28.

Di Toro, D.M. 1985. A particle interaction model of reversible organic chemical sorption. *Chemosphere* 14:1503–38.

Dieulin, A., G. Matheron, and G. De Marsily. 1981. Growth of the dispersion coefficient with the mean traveled distance in porous media. *The Science of the Total Environment* 21:319–28.

Donigian, A.S., and L.A. Mulkey. 1987. *Stream—An Exposure Assessment Methodology for Agricultural Pesticide Runoff, Proceedings: US/USSR Symposium on Fate of Pesticides and Chemicals in the Environment.* Iowa City: University of Iowa. 32 pp.

Duo-Sen, L., and Z. Shui-Ming. 1986. Kinetic model for degradative processes of pesticides in soil. *Ecological Modeling* 37:131–38.

Dzantro, E.K., and A.S. Felsot. 1991. Microbial responses to large concentrations of herbicides in soil. *Environ. Toxic. & Chem.* 10:649–55.

Ehteshami, M., R.C. Peralta, H. Eisele, H. Deer, and T. Tindall. 1991. Assessing pesticide contamination to ground water: A rapid approach, *Ground Water* November–December:862–68.

Ekler, Z. 1988. Behavior of thiocarbamate herbicides in soils: adsorption and volatilization, *Pestic. Sci.* 22:145–57.

Elabd, H., W.A. Jury, and M.M. Cliath. 1986. Spatial variability of pesticide adsorption parameters. *Environ. Sci. Technol.* 20:256–60.

Emmerich, W.E., D.A. Woolhiser, and E.D. Shirley. 1989. Comparison of lumped and distributed models for chemical transport by surface runoff. *J. Environ. Qual.* 18:120–26.

Eschenroeder, A. 1983. The role of multimedia fate models in chemical risk assessment. In *Fate of Chemicals in the Environment*, ACS Symposium Series, pp. 89–104.

Everts, C.J., and R.S. Kanwar. 1988. Quantifying preferential flow to a tile line with tracers. Presented at 1988 Winter Meeting of the American Society of Agricultural Engineers, Chicago, 13–16 Dec. 1988.

Fadayomi, O., and G.F. Warren. 1977. Adsorption, desorption, and leaching of nitrofen and oxyfluorfen. *Weed Science* 25:97–100.

Frehse, H. and J.P.E. Anderson. 1983. Pesticide residues in soil—problems between concept and concern. In *Pest. Chem. Human Welfare Environ.*, Vol. 4, ed. Miyamato and Kearney,

Fuesler, T.P., and M.K. Hanafey. 1990. Effect of moisture on chlorimuron degradation in soil. *Weed Science* 38:256–61.

Galiulin, R.V., Y.B. Mironyenko, Y.A. Pachepsky, and M.S. Sokolov. 1984. Mathematical modeling of the rate of herbicide dissipation in soil. *Agrochimia* 6:92–100.

Gamerdinger, A.P., A.T. Lemley, and R.J. Wagenet. 1991. Nonequilibrium sorption and degradation of three 2-chloro-s-triazine herbicides in soil-water systems, *J. Environ. Qual.* 20:815–22.

Gamerdinger, A.P., R.J. Wagenet, and M.Th. Van Genuchten. 1990. Application of two-site/two-region models for studying simultaneous nonequilibrium transport and degradation of pesticides. *Soil Sci. Soc. Am. J.* 54:957–63.

Gardner, W.R., W.A. Jury and J. Knight. 1974. Water uptake by vegetation. *J. Agronomy* 66:443–56.

Gill, W. N., and R. Sankarasubramanian. 1970. Exact analysis of unsteady convective diffusion. *Proc. Roy. Soc. Lond.* A316:341–50.

Gish, T.J., C.S. Helling, and P.C. Kearney. 1986. *Simultaneous Leaching of Bromide and Atrazine Under Field Conditions, Proc. Agricultural Impacts on Ground Water,* 1986, Aug 11–13, Omaha, Nebr. NWWA, pp. 286–97.

Glotfelty, D.E., A.W. Taylor, B.C. Turner, and W.H. Zoller. 1984. Volatilization of surface-applied pesticides from fallow soil. *J. Agri. Food Chem.* 32:638–43.

Glotfelty, D.E., G.H. Williams, H.P. Freeman, and M.M. Leech. 1989. Regional atmospheric transport and depositions of pesticides in Maryland. *J. Agri Food Chem.* 37:331–39.

Glotfelty, D.E., J.N. Seiber, and L.A. Liljedahl. 1987. Pesticides in fog. *Nature* 325:602–605.

Glotfelty, D.E., M.S. Majewski, and J.M. Seiber. 1990. Distribution of several organophosphorous insecticides and their oxygen analogs in a foggy atmosphere. *Environ. Sci. Technol.* 24:353–57.

Golovleva, L.A., N. Aharonson, R. Greenhalgh, N. Sethunathan, and J.W. Vonk. 1990. The role and limitations of microorganisms in the conversion of xenobiotics. *Pure & Appl. Chem.* 62:351–64.

Goring, C.A.I., and J.W. Hamaker. 1971. The degradation and movement of picloram in soil and water. *Down to Earth* 27:12–15.

Goss, D., and R.D. Wauchope. 1990. The SCS/ARS/CES Pesticide Properties Database: II using it with soils data in a screening procedure. In *Pesticides in the Next Decade: The Challenges Ahead, Proceedings of the Third National Research Conference on Pesticides,* November 8–9, 1990, ed. D. L. Weigmann, pp. 471–93.

Gustafson, D.I., and L.R. Holden. 1990. Nonlinear pesticide dissipation in soil: a new model based on spatial variability. *Environ. Sci. Technol.* 24:1032–38

Gustafson, D.I. 1988a. Modeling root zone dispersion: a comedy of error functions. *Chem. Eng. Comm.* 73:77–94.

Gustafson, D.I. 1988b. Accuracy of predictive water quality models—comparison with measured surface water concentrations of crop chemicals in northern Ohio. *Environ. Toxic. Chem.* 7:261–62.

Gustafson, D.I. 1989. Groundwater ubiquity score: a simple method for assessing pesticide leachability. *Environ. Toxic. Chem.* 8:339–57.

Gustafson, D.I. 1990. Field calibration of surface: a model of agricultural chemicals in surface waters. *J. Environ. Sci. & Health* B25:665–87.

Hamaker, J.W., C.R. Youngson, and C.A.I. Goring. 1967. Prediction of the persistence and activity of tordon herbicide in soils under field conditions. *Down to Earth* Fall:30–39.

Harder, H.W., E.C. Christioansen, J.R. Matthews, and T.F. Bidleman. 1980. Rainfall input of toxaphene to a South Carolina estuary. *Estuaries* 3:142–47.

Harris, C.R., R.A. Chapman, J.H. Tolman, P. Moy, K. Henning, and C. Harris. 1988. A comparison of the persistence in a clay loam of single and repeated annual applications of seven granular insecticides. *J. Environ. Sci. Health* B23:1–23.

Harvey, R.G. 1987. Herbicide dissipation from soils with different herbicide use histories. *Weed Science* 35:583–89.

Heathman, G.C., L.R. Ahuja, and O.R. Lehman. 1985. The transfer of soil surface-applied chemicals to runoff. *Trans. ASAE* 28:1909–20.

Hill, B.D. and G.B. Schaalje. 1985. A two-compartment model for the dissipation of deltamethrin on soil. *J. Agric. Food Chem.* 33:1001–1006.

Hillel, D. 1980. *Fundamentals of Soil Physics,* New York: Academic Press. pp. 200–201.

Hoag, D.L., and A.G. Hornsby. 1991. Coupling groundwater contamination to economic returns when applying farm pesticides. *Dare*:91–08, North Carolina State University, Raleigh N.C. 26 pp.

Hoag, D.L. 1990. *Reducing Agrichemical Risk in a Competitive Market System, Proceedings of the Annual Meeting of the American Association for the Advancement of Science,* New Orleans, 1990.

Hyzak, D.L., and R.L. Zimdahl. 1974. Rate of degradation of metribuzin and two analogs in soil. *Weed Science* 22:75–79.

Jafvert, C.T. 1990. Sorption of organic acid compounds to sediments: initial model development. *Environ. Toxic. & Chem.* 9:1259–68.

Jones, R.L., G.W. Black, and T.L. Estes. 1986. Comparison of computer model predictions with unsaturated zone field data for aldicarb and aldoxycarb. *Environ. Toxic. Chem.* 5:1027–37.

Jury, W.A., and L.H. Stolzy. 1982. A field test of the transfer function model for predicting solute transport. *Water Resources Research* 18:369–75.

Jury, W.A., 1982. Simulation of solute transport using a transport function model, *Water Resources Research* 18:363–68.

Jury, W.A., D.D. Focht, and W.J. Farmer. 1987. Evaluation of pesticide groundwater pollution potential form standard indices of soil-chemical adsorption and biodegradation. *J. Environ. Qual.* 16:422–28.

Jury, W.A., J.S. Dyson, and G.L. Butters. 1990. Transfer function model of field-scale solute transport under transient water flow. *Soil Sci. Soc. Am. J.* 54:327–32.

Kalouskova, N. 1987. Interaction of humic acids with atrazine. *J. Environ. Sci. Health B* 22:113–23.

Kaufman, D.D., and J. Blake. 1972. Microbial degradation of several acetamide, acylanilide, carbamate, toluidine and urea pesticides. *Soil Biol. Biochem.* 5:297–308.

Kirby, J.M. 1985. A note on the use of a simple numerical model for vertical, unsaturated fluid flow. *Soil Science* 139:462–67.

Leake, C.R., D.J. Arnold, S.E. Newby, and L. Somerville. 1987. *Benazolin-Ethyl—A Case Study of Herbicide Degradation and Leaching* (1987 BCPC-Weeds), pp. 577–83.

Lee, L.S., P.S.C. Rao, and M.L. Brusseau. 1991. Nonequilbrium sorption and transport of neutral and ionized chlorophenols. *Environ. Sci. Technol.* 25:722–29.

Leonard, R.A., and W.G. Knisel. 1988. Evaluating groundwater contamination potential from herbicide use. *Weed Technology* 2:207–16.

Leonard, R.A., A. Shirmohammadi, A.W. Johnson, and L.R. Marti. 1988. Pesticide transport in shallow groundwater. *Trans. ASAE* 31:776–88.

Ligocki, M.P., C. Leuenberge, and J.F. Pankow. 1985. Trace organic compounds in rain—II. gas scavenging of neutral organic compounds. *Atmos. Environ.* 19:1609–17.

Lim, S.U., J.K. Lee, and K.H. Han. 1977. Behaviors of some pesticides in soils (part

I): adsorption of the herbicides atrazine and alachlor. *Hanguk Nonghwa Hakhoe Chi* 20:310–16.

Lindstrom, F.T., and W.T. Piver. 1985. A mathematical model for the transport and fate of organic chemicals in unsaturated/saturated soils. *Environ. Health Perspec.* 60:11–28.

Liu, C.C.K., K. Loague, and J.-S. Feng. 1991. Fluid flow and solute transport processes in unsaturated heterogeneous soils: preliminary numerical experiments. *J. Contam. Hydrol.* 7:261–83.

Lorber, M.N., and L.A. Mulkey. 1982. An evaluation of three pesticide runoff loading models. *J. Environ. Qual.* 11:519–29.

Lym, R. F., and O. R. Swenson. 1991. Sulfometuron persistence and movement in soil and water in North Dakota. *J. Environ. Qual.* 20:209–15.

Mackay, D., and B. Powers. 1987. Sorption of hydrophobic chemicals from water: a hypothesis for the mechanism of the particle concentration effect. *Chemosphere* 16:745–57.

Mackay, D., and S. Paterson. 1981. Calculating fugacity. *Environ. Sci. Tech.* 15:1006–14.

Mackay, D., and S. Paterson. 1982. Fugacity revisited. *Environ. Sci. Tech.* 16:654a–660a.

Mackay, D., and S. Paterson. 1984. Spatial concentration distributions. *Environ. Sci. Tech.* 207a–214a

Mackay, D., and S. Paterson. 1991. Evaluating the multimedia fate of organic chemicals: a level Iii fugacity model. *Environ. Sci. Technol.* 25:427–36.

Mackay, D. 1979. Finding fugacity feasible. *Environ. Sci. Tech.* 13:1218–23.

Maguire, R.J. 1991. Kinetics of pesticide volatilization from the surface of water. *J. Agric. Food Chem.* 39:1674–78.

Majewski, M.S., D.E. Glotfelty, K.T.P. U, and J.N. Seiber. 1990. A field comparison of several methods for measuring pesticide evaporation rates from soil. *Environ. Sci. Technol.* 24:1490–97.

Majewski, M.S., M.M. McChesney, and J.N. Seiber. 1991. A field comparison of two methods for measuring DCPA soil evaporation rates. *Environ. Toxic. & Chem.* 10:301–11.

McConnell, J.S., and L.R. Hossner. 1985. Ph-dependent adsorption isotherms of glyphosate. *J. Agric. Food Chem.* 33:1075–78.

McKone, T.E. 1987. Human exposure to volatile organic compounds in household tap water: the indoor inhalation pathway. *Environ. Sci. Tech.* 21:1194–1201.

Melancon, S.M., J.E. Pollard, and S.C. Hern. 1986. Evaluation of SESOIL, PRZM and PESTAN in a laboratory column leaching experiment. *Environ. Tox. Chem.* 5:865–78.

Mersie, W., and C.L. Foy. 1988. Adsorption, desorption, and mobility of chlorsulfuron in soils. *J. Agric. Food Chem.* 34:89–92.

Mills, W.C., and R.A. Leonard. 1984. Pesticide pollution probabilities. *Trans. ASAE,* 27:1704–10.

Nash, R.G. 1983. Comparative volatilization and dissipation rates of several pesticides from soil. *J. Agric. Food Chem.* 31:210–17.

Newell, C.J., L.P. Hopkins, and P.B. Bedient. 1990. A hydrogeologic database for ground-water modeling. *Ground Water* 28:703–14.

Nicholls, P.H., and A.A. Evans. 1991. Sorption of ionizable organic compounds by field soils. *Pestic. Sci.* 33:319–45.

Nicholls, P.H. 1988. Factors influencing entry of pesticides into soil water. *Pestic. Sci.* 22:123–37.

Nicholls, P.H., R.H. Bromilow, and T.M. Addiscott. 1982. Measured and simulated behavior of fluometuron, aldoxycarb and chloride ion in a fallow structured soil. *Pesticide Science* 13:475–83.

Nigg, H.N., and J.C. Allen. 1979. A comparison of time and time-weather models for predicting parathion disappearance under California conditions. *Environ. Sci. Tech.* 13:231–33.

Nose, K. 1986. A multi-site decay model of pesticide in soil. *J. Pestic. Sci.* 12:505–508.

O'Connor, G.A., M.T. Van Genuchten, and P.J. Wierenga. 1976. Predicting 2,4,5-T movement in soil columns. *J. Environ. Qual.* 5:375–78.

Obrigawitch, T., R.G. Wilson, A.R. Martin, and F.W. Roth. 1982. The influence of temperature, moisture, and prior EPTC application on the degradation of EPTC in soils. *Weed Science* 30:175–81.

Ou, L.-T., P.S.C. Rao, K.S.V. Edvardsson, R.E. Jessup, A.G. Hornsby, and R.L. Jones. 1988. Aldicarb degradation in sandy soils from different depths. *Pestic. Sci.* 23:1–12.

Padilla, F., P. Lafrance, C. Robert, and J.-P. Villeneuve. 1988. Modeling the transport and the fate of pesticides in the unsaturated zone considering temperature effects. *Ecological Modeling* 44:73–88.

Pignatello, J.J., and L.Q. Huang. 1991. Sorptive reversibility of atrazine and metolachlor residues in field soil samples. *J. Environ. Qual.* 20:222–28.

Rao, P.S.C., and R.E. Jessup. 1983. Sorption and movement of pesticides and other toxic organic substances in soil. *Soil Sci. Soc. of Am. Spec. Publ.* 11:183–201.

Rao, P.S.C., and R.J. Wagenet. 1985. Spatial variability of pesticides in field soils: methods for data analysis and sampling. *Weed Science,* 33 (Suppl. 2):18–24.

Rao, P.S.C., K.S.V. Edvardsson, L.T. Ou, P. Nkedi-Kizza, and A.G. Hornsby. 1986. Spatial variability of pesticide sorption and degradation parameters. In *Evaluation of Pesticides in Groundwater,* Washington D.C.: ACS. 100–115.

Reinert, K.H., P.M. Rocchio, and J.H. Rodgers. 1987. Parameterization of predictive fate models: a case study. *Environ. Toxic. Chem.* 6:99–104.

Richards, R.P., J.W. Kramer, D.B. Baker, and K.A. Krieger. 1987. Pesticides in rainwater in the northeastern united states. *Nature* 327:129–31.

Sadeghi, A.M., D.E. Kissel, and M.L. Cabrera. 1989. Estimating molecular diffusion coefficients of urea in unsaturated soil. *Soil Sci. Soc. Am. J.* 53:15–18.

Sanchez-Camazano, M., and M.J. Sanchez-Martin. 1988. Influence of soil characteristics on the adsorption of pirimicarb. *Environ. Tox. Chem.* 7:559–64.

Sanders, P.F., and J.N. Seiber. 1984. Organophosphorous pesticide volatilization. In *Treatment and Disposal of Pesticide Wastes,* ACS Symposium Series.

Sato, T., S. Kohnosu, and J.F. Hartwig. 1987. Adsorption of butachlor to soils. *J. Agric. Food Chem.* 35:397–402.

Sethi, R.K., and S.L. Chopra. 1975. Adsorption, degradation, and leaching of alachlor in some soils. *J. Indian Soc. Soil Sci.* 24:184–94.

Shirmohammadi, A., W.L. Magette, R.B. Brinsfield, and K. Staver. 1989. Ground Water loading of pesticides in the Atlantic coastal plain. *GWMR* Fall:141–48.

Sips, R. 1950. On the structure of a catalyst surface. *J. Chem. Phys.* 18:1024–26.

Smith, A.E., and A. Walker. 1977. A quantitative study of asulam persistence in soil. *Pestic. Sci.* 8:449–56.

Smith, A.E., and A. Walker. 1989. Prediction of the persistence of the triazine herbicides atrazine, cyanazine, and metribuzin in regina heavy clay. *Can. J. Soil Sci.* 69:587–95.

Smith, S.J., and R.J. Davis. 1974. Relative movement of bromide and nitrate through soils. *J. Environ. Qual.* 3:152–55.

Smith, S.J., L.R. Ahuja, and J.D. Ross. 1984. Leaching of a soluble chemical under field crop conditions. *Soil Sci. Soc. Am. J.* 48:252–58.

Somasundaram, L., J.R. Coats, K.D. Racke, and V.M. Shanbhag. 1991. Mobility of pesticides and their hydrolysis metabolites in soil. *Environ. Toxic. & Chem.* 10:185–94.

Southwood, J.M., R.C. Harris, and D. Mackay. 1989. Modeling the fate of chemicals in an aquatic environment: the use of computer spreadsheet and graphics software. *Environ. Toxic. & Chem.* 8:987–96.

Stamper, J.H., H.N. Nigg, and J.C. Allen. 1979. Organophosphate insecticide disappearance from leaf surfaces: an alternative to first-order kinetics. *Environ. Sci. Tech.* 13:1402–1406.

Starr, J.L., and D.E. Glotfelty. 1990. Atrazine and bromide movement through a silt loam soil. *J. Environ. Qual.* 19:552–58.

Stauffer, T.B., W.G. Macintyre, and D.C. Wickman. 1989. Sorption of nonpolar organic chemicals on low-carbon-content aquifer materials. *Environ. Toxic. & Chem.* 8:845–52.

Steenhuis, T.S., and L.M. Naylor. 1987. A screening method for preliminary assessment of risk to groundwater from land-applied chemicals. *J. Contam. Hydrol.* 1:395–406.

Stenstrom, J. 1989. Quantitative assessment of herbicide decomposition at different initial concentrations. *Toxicity Assessment* 4:53–70.

Sundaram, K.M.S. 1991. Fate and short-term persistence of permethrin insecticide injected in a northern Ontario (Canada) headwater stream. *Pestic. Sci.* 31:281–94.

Taylor, A.W., D.E. Glotfelty, B.L. Glass, H.P. Freeman, and W.M. Edwards. 1976. Volatilization of dieldrin and heptachlor from a maize field. *J. Agric. Food Chem.* 24:625–31.

Taylor, G.I. 1953. Dispersion of soluble matter in solvent flowing slowly through a tube. *Proc. Roy. Soc. Lond.* 219a:186–89.

Vaccari, D.A., and M. Kaouris. 1988. A model for irreversible adsorption hysteresis. *J. Environ. Sci. Health* A23:797–822.

Van Genuchten, M.Th., and R.J. Wagenet. 1989. Two-site/two-region models for pesticide transport and degradation: theoretical development and analytical solutions. *Soil Sci. Soc. Am. J.* 53:1303–10.

Wagenet, R.J., J.L. Hutson, and J.W. Biggar. 1989. Simulating the fate of a volatile pesticide in unsaturated soil: a case study with DBCP. *J. Environ. Qual.* 18:78–84.

Walker, A. 1976. Simulation of herbicide persistence in soil. *Pestic. Sci.* 7:41–49.

Walker, A., and S.J. Welch. 1991. Enhanced degradation of some soil-applied herbicides. *Weed Research* 31:49–57.

Walker, W.W., C.R. Cripe, P.H. Pritchard, and A.W. Bourquin. 1988. Biological and abiotic degradation of xenobiotic compounds in vitro estuarine water and sediment/water systems. *Chemosphere* 17:2255–70.

Wauchope, R.D., T.M. Buttler, A.G. Hornsby, P.W.M. Augustijn-Beckers, and J.P. Burt. 1992. The SCS/ARS/CES pesticide properties database for environmental decision-making. *Reviews of Environmental Contamination and Toxicology* 123:1–164.

Wehtje, G., R. Dickens, J.W. Wilcut, and B.F. Hajek. 1987. Sorption and mobility of sulfometuron and imazapyr in five Alabama soils. *Weed Science* 35:858–64.

Wood, L.S., H.D. Scott, D.B. Marx, and T.L. Lavy. 1987. Variability in sorption coefficients of metolachlor on a captina silt loam. *J. Environ. Qual.* 16:251–56.

Wu, T.C., Y.S. Wang, and Y.L. Chen. 1991. Degradation of isouron in three upland soils in Taiwan. *J. Pesticide Sci.* 16:195–200.

Zimdahl, R.J., and S.K. Clark. 1982. Degradation of three acetanilide herbicides in soil. *Weed Science* 30:545–48.

Zins, A.B., D.L. Wyse, and W.C. Koskinen. 1991. Effect of alfalfa (medicago sativa) roots on movement of atrazine and alachlor through soil. *Weed Science* 39:262–69.

Zurmuhl, T., W. Durner, and R. Herrmann. 1991. Transport of phthalate-esters in undisturbed and unsaturated soil columns. *J. Contam. Hydrol.* 8:111–33.

III

What's Being Done About It?

5

Regulations Controlling Pesticides in Drinking Water

Needless to say, the occurrence of pesticides in drinking water, whether a result of accidents or natural processes following proper application in an agricultural situation, has initiated a complete reexamination of the procedures that had been used to regulate, apply, and develop pesticides. The final three chapters of this book will describe the changes in these three areas. In describing the regulatory strategies devised to manage and minimize the occurrence of pesticides in drinking water supplies, an effort will be made to discuss national (U. S.) regulations, international (mainly European Community) efforts, and individual state efforts.

TIERED REGULATORY STRATEGIES

Because not all pesticides share the same potential for reaching drinking water, and because they would cause different levels of concern if they were present, most regulatory strategies have a tiered framework. As individual triggers or thresholds are met by the pesticide's perceived contamination potential, additional regulatory requirements are imposed. While varying slightly worldwide, the basic progression is:

1. Laboratory studies;
2. Field studies; and
3. Large-scale monitoring and modeling.

The laboratory studies are useful for isolating individual processes, such as sorption, biodegradation, or volatilization. Field experiments reveal the integrated effect of the various phenomena and how they are affected by soil type, climate, and cultural practice. Besides these direct experimental approaches, several mathematical models of the leaching process are used in both the intermediate and final stages of this regulatory process.

The laboratory studies designed to measure the potential for drinking water contamination include determinations of the key thermodynamic equilibrium properties of the molecule: water solubility, vapor pressure, pKa, pKb, and partition coefficients between both soil/water and octanol/water. Besides these equilibrium values, certain kinetic parameters are determined, including rates of aqueous hydrolysis, photolysis (on soil or in water), and metabolism in natural soils and water under both aerobic and anaerobic conditions. Virtually all of these laboratory studies are performed using ^{14}C radiolabeled material, which allows for the identification of pesticide degradates. The biological activity of these degradates is assessed in order to determine whether they should be considered when environmental assessments are made.

These laboratory studies sometimes include special studies geared toward directly measuring the mobility of pesticides and/or their degradates in soil. Generally called soil column studies, these experiments rely upon the use of small diameter ($\sim$10 cm) cylinders of soil (30 cm in length). A small quantity of pesticide, usually radiolabeled, is applied to the top surface of the soil and water is applied. If done immediately, such experiments measure only the mobility of the parent compound, but if a period of time is allowed to elapse before applying the large quantity of water, such an experiment is usually called an "aged-leaching study." The protocols for such experiments vary widely, but after a specified period of time (generally about one month), a large quantity of water is applied to the soil column and the amount of applied radioactivity present in the leachate is determined. While not directly applicable to what might occur in the field, the results of such a test are generally taken as a relative measure of mobility of both the parent pesticide and its daughter products in soil.

After the mobility and persistence of the pesticide and its degradates have been determined under laboratory conditions, the first field studies are conducted. These field experiments generally represent the second tier of the assessment scheme. In some countries, these field studies are not needed when the pesticide is shown to be sufficiently immobile and readily degraded. In most countries, however, field studies are required for all compounds. This is certainly the case in the United States. As required by the EPA, field dissipation studies are designed to measure, under actual conditions of use, the dissipation rate of the pesticide and its degradates. In addition, because sampling is performed to at least 90 centimeters in depth, with samples analyzed in 15-centimeter increments, definitive information on the mobility of the materials under field conditions is obtained.

When field studies suggest or show that the pesticide occurs in drinking water supplies, then large-scale monitoring is required if the pesticide is already widely used. If the pesticide is still in development, such a finding would generally require a much more detailed and costly experiment to assess whether the concentrations likely to occur in drinking water pose a human health threat. The design of such monitoring studies in the United States is discussed in greater detail below.

Before launching into a specific discussion of regulatory strategies worldwide, some description of the newer experimental techniques to measure pesticide transport and fate will be given. Most of these methods fall between levels 1 and 2 of the tiered regulatory scheme. They are generally experiments that are done in the field, but under much more controlled conditions, resembling the laboratory. As one example of an experiment, Wauchope, Williams, and Marti (1990) have developed tilted-bed simulators as a convenient and inexpensive method for measuring the runoff potential of pesticides from agricultural fields. Another similarly innovative device has been developed for the *in situ* measurement of retardation factors in aquifers (Gillham, Robin, and Ptacek 1991).

The use of tile-line effluent studies as a field tool to evaluate the potential impacts of pesticides on groundwater quality has been described (Hallberg, Baker, and Randall 1987). Monitoring of tile-line effluents in the Big Spring basin of northeastern Iowa suggested that water quality of these streams is similar to the groundwater there. Whether these data can be extrapolated to non-karst areas is unclear.

Model ecosystems or microcosms have been used to assess the impact of pesticides on aquatic life (Yu et al. 1975; Miyamoto et al. 1985). They are said to be useful screening tools for assessing the environmental fate of pesticides, but they have several limitations:

- The number of organisms and environmental parameters measured is limited.
- Food chain organisms are not truly representative of the natural environment.
- Lower trophic organisms are frequently consumed in too short a time period.
- Component organisms are not kept under their physiological optimum conditions.

There are compromises that need to be made in the use of these systems in order to account for as many natural processes as possible, while retaining the desirable aspects of reproducibility and ease of use.

Of all the new experimental techniques involved in the regulation of pesticides in drinking water, the one receiving the most attention is soil lysimeters. These devices, first used to describe the leaching of soil solutes in the late-nineteenth century, are generally composed of an intact cylindrical soil core of varying diameter but almost always with a length of 1 meter. Radiolabeled pesticides are usually applied, and the lysimeters are typically kept in the field. The advantages of such systems are that all water can be collected, and the use of radiolabeled materials allows reliable determination of whether the pesticide or any of its degradates has a measurable leaching potential under field conditions in "real" soils. The disadvantage is the possibility that physical imperfections in the lysimeters (cracks along the side) may give spurious results. Nevertheless, they have become widely used, especially in the European Community, as an integral part of the registration process. Several recent examples of their use are described here.

The leaching potential and decomposition of clopyralid in Swedish soils under field conditions has been studied using field lysimeters (Bergstrom et al. 1991). Lysimeters were kept in PVC containers with an inner diameter of 0.295 meters and a length of 1.18 meters. Normal and double doses of the compound, 120 and 240 g/ha, were used, and additional irrigation was supplied to certain lysimeters. Traces of clopyralid were found in two water samples that occurred in a very heavy rain immediately following an extended dry period. This suggests that the compound leached as a result of macropores rather than leaching, as would be suggested by the convective-dispersion equation presented in Chapter 4. The authors theorized that clay soils, which have a greater tendency to exhibit macropore features, might be more susceptible to leaching than sandy soils are. This observation, while perhaps consistent with the Swedish data, is not consistent with the main body of pesticide in drinking water data presented elsewhere in this text. Despite the occurrence of residues in the single sample, the authors went on to conclude that their results showed clopyralid would be unlikely to show significant leaching tendencies under Swedish conditions, both in clay soils and in light-textured (sandy) soils.

Good agreement between conventional field leaching studies and lysimeter work has been reported for studies involving bromide movement (Saffinga et al. 1984), metamitron and methabenzthiazuron (Kubiak et al. 1988), and one experimental insecticide (Kordel et al. 1988; Herrchen et al. 1990), but considerably more data must be obtained in order to conclude that the two are truly in agreement.

Two methods for the collection of intact soil cores for use in lysimeters were described by Leake (1991). In one technique, a circular PVC plastic pipe (0.1 meters i.d. × 1 meter) with a tapered end was driven into the ground with a hammer drill, and the intact core was simply winched out of the ground. Larger lysimeters (0.5 meters i.d. × 1 meter) were collected by attaching a steel cutting ring that provided mechanical excavation of an annulus around the PVC pipe as it was pressed down into the soil. Further excavation of the soil surrounding the inserted device was required in order to remove it intact.

Another researcher who has spent considerable effort investigating the utility of field lysimeters in assessing the potential for groundwater contamination by pesticides is Bruce Bowman of Agriculture Canada. In a series of papers (Bowman 1988, 1989a, 1989b, 1990, 1991a, 1991b), he has discussed results obtained with packed soil lysimeters of a particular geometry and type (stainless steel cylinders 0.15 meters in diameter and 0.7 meters long) buried in an outdoor sandbox. In these systems he has generated extensive data on the effect of soil type, irrigation, pesticide properties, and pesticide formulation on the potential for pesticide leaching. The question of whether Bowman's use of packed rather than intact soil columns has caused some to question his results, but the relative trends he has observed are almost certainly in agreement with field behavior in intact sandy soils.

A new term for these smaller soil lysimeters was coined by one group of researchers (Fermanich, Daniel, and Lowery 1991), who called their system a micro-lysimeter soil column. A 1-meter length of 0.2-meter diameter aluminum irrigation pipe was slowly pressed into the soil profile with a backhoe to a depth of approximately 90 centimeters. After insertion, a hole was dug adjacent to the cylinder and the intact soil column was extracted with minimal disturbance. Once returned to the laboratory, the base of each micro-lysimeter was inserted into an acrylic vacuum chamber fitted with a porous stainless steel plate. Window glazing compound was used to give an airtight seal between the plate and vacuum chamber. A 0.5-centimeter layer of fine sand was applied to assure complete hydraulic contact between the soil column and steel plate. Field drainage conditions were simulated by applying a constant suction of 9 kPa.

A similar apparatus was used to study the movement of metamitron and chloridazon in Spain (Goicolea et al. 1991). Besides having utility in the study of pesticide fate and transport in soil, lysimeters can be used to study the effect of genetically engineered organisms on nutrient dynamics (Fredrickson et al. 1990).

The trend in regulatory strategies worldwide is toward the wider use of such controlled systems utilizing radiolabeled material to ensure a complete mass balance for the pesticide. The problem still to be resolved is how to interpret the results of these experiments, extrapolate them to likely behavior under field conditions, and use that information in developing a logical overall regulatory strategy. One integral part of the overall strategy is assessing whether the observed or predicted concentrations pose a threat to human health. That is the subject of the next section.

HEALTH-BASED DRINKING WATER
STANDARDS FOR PESTICIDES

Under the Safe Drinking Water Act in the United States, the EPA has been assigned the responsibility for establishing maximum contaminant levels (MCLs) for certain toxic chemicals, including pesticides. All public water supply systems (using either ground or surface water sources) must verify, generally through quarterly sampling, that the water they supply their customers does not contain individual pesticides at levels exceeding these concentrations. If exceedence does occur, the water must be treated to bring levels down below the relevant MCLs. In order to set MCLs, the EPA has developed a procedure in which a maximum contaminant level goal (MCLG) is first assigned for the toxic chemical based solely on health effects. This goal is assigned based on observations of toxicity either in humans or, more commonly, in representative mammals (rats, mice, rabbits, etc.).

Toxicity

Taken in sufficient quantities, all chemicals, including the naturally occurring materials in the food we eat, are toxic. According to the famous quote, it is the dose that makes the poison. For all pesticides, there is a standard battery of toxicity testing that must be undertaken in order to register the compound for its use. This testing is divided primarily into two categories: acute and chronic. In acute testing, single doses of the chemical are administered to animals at a range of levels to determine the lethal dose. The interpolated dose at which one-half the test animals are killed is called the LD50. This dose is generally quoted on a (mg pesticide/kg body weight) basis, in order to facilitate extrapolation to other species.

In addition to such acute exposures, chronic testing is performed in which the test animals are fed diets containing various pre-dosed levels of the test chemical. The levels used to dose the food are usually in the ppm range, and the results of the testing are generally expressed on a (mg pesticide/kg body weight/day) basis, again in order to facilitate extrapolation to other species.

For example, a long-term feeding study might be performed by feeding rats various doses of a pesticide corresponding to doses of 0, 1, 5, 10, and 100 mg/kg/day. Let us assume that statistically significant toxic effects on the liver are observed at 10 mg/kg/day and above, but that no measurable or significant effects are found at the two lower doses. The no effect level for the study (or NOEL) would be set at 5 mg/kg/day. In extrapolating this NOEL to humans, in the absence of any more detailed information, a safety or uncertainty factor of 100 would be employed, leading to a reference dose (often denoted RfD) of 0.05 mg/kg/day. If the pesticide in question is used on food crops, it is then typically assumed that 20 percent of one's exposure comes from the consumption of water, leading to a target dose in drinking water of 0.01 mg/kg/day. For a 60-kg adult consuming 2 L of water per day, this works out to an MCLG of $0.01*60/2 = 0.3$ mg/L or 300 $\mu g/L$. If it is technically feasible to measure such concentrations reliably in drinking water and there are available technologies to lower concentrations to this level, then the MCL will be set equal to this MCLG. However, it is sometimes the case that the MCLG calculated as in the above example is at a level well below what can be reliably determined using existing analytical methodologies or at a level that available treatment technologies are unable to attain. In such a case, the MCL will be set at the lowest level feasible, using the existing technology for measuring the concentration of the pesticide or for removing it. This would typically result in a slightly lower safety factor with regard to the effects of the pesticide on human health, and it will need to be verified that the overall benefits of continued use of the pesticide to society outweigh these increased risks.

The example given above refers to a noncarcinogenic toxic effect at the high dosage level. If tumors are observed in more than one species of mammal, the MCLG is arbitrarily set to zero in most cases. This approach is now in question, and there is considerable debate in the toxicology community on the appropriate way to manage the rodent bioassay infor- mation (Lave et al. 1988). As these authors have pointed out, only 26 chemicals or groups of chemicals of the commonly used 60,000 materials have been shown to have definitive evidence of human carcinogenicity, yet fully 65 percent of the 800 that have undergone the typical rodent bioassay are positively linked to cancer. Clearly, much of this is due to the very high rates of exposure employed in the rodent bioassay in relation to the very low exposures experienced by most of society. There is also the fact that carcinogenic potentials are not always exhibited across species. The ob- served concordance between rats and mice, for instance, is only about 70 percent. The value of running an extremely expensive (over $1 million) test to measure cancer-causing potential in rats and mice is obviously questionable when the accuracy in simple extrapolation is so uncertain. The value of such continued testing may not be the direct impact of such experiments in providing protection to the public, but in providing data to pharmacokinetics models and other techniques to improve the extrapola- tion of test results.

Removal Technologies

Man has always drunk water and has always had discriminating tastes. Various methods have been handed down through the ages for improving the taste of local waters, and in more recent years the technology of converting raw surface water into a potable supply of acceptable quality has become increasingly complex. The presence of organic materials in drinking water has always been distasteful, and the methods to remove such contaminants are practically identical to those used to remove pesticides from drinking water. For virtually all pesticides, the method of choice is granular activated carbon (GAC) filters. These can be installed at relatively low cost (several hundred dollars) on private wells, with replacement of the filters every couple of years.

A rather complete discussion of current removal technologies is available in a recent paper by EPA personnel (Goodrich, Lykins, and Clark 1991). They report that the preferred method of removal is granular activated carbon, although air stripping can be helpful for very volatile materials such as DBCP and EDB. Nitrate and nitrite ion, being inorganic, do not sorb to carbon and the more expensive technologies, ion exchange and reverse osmosis, are the best available techniques for these contaminants.

The effectiveness of photolytic ozonation in reducing the concentrations of alachlor in drinking water has been demonstrated (Somich et al. 1988). One exciting technique for removing such materials was recently reported (BPI 1991). The Solar Research Institute in Golden, Colorado, has found that pumping polluted water through long, narrow glass tubes in curved glass troughs can cause photolytic degradation of the organic chemicals.

For most pesticides, removal techniques are routinely available and will not be the limiting factor in setting the MCL as close as feasible to the MCLG. The more common problem is the analytical difficulty in reliably quantifying through analytical methods the rather low concentrations (sometimes below $1 \mu g/L$) at which the MCLG has been set.

Besides the MCLs and MCLGs mandated under the Safe Drinking Water Act, the EPA establishes Health Advisory Levels (HALs) based on various time periods of consumption (one-day, ten-day, lifetime) that are perceived to be safe according to the available toxicity information for the pesticide. Unlike MCLs, the HALs do not carry with them any requirement for public drinking water systems to ensure they are in compliance, but they do serve a useful purpose in communicating whether an observed level of a pesticide in drinking water has negligible risk associated with it.

Appendix 1 contains a list of several pesticides now in use worldwide, with an indication of the health-based maximum acceptable concentrations that have been established by five countries. Besides listing HALs and MCLs for pesticides in the United States, similar concentrations are listed for Australia, Canada, Germany, and Great Britain. As should be obvious from the concentrations given in this appendix, there is generally very

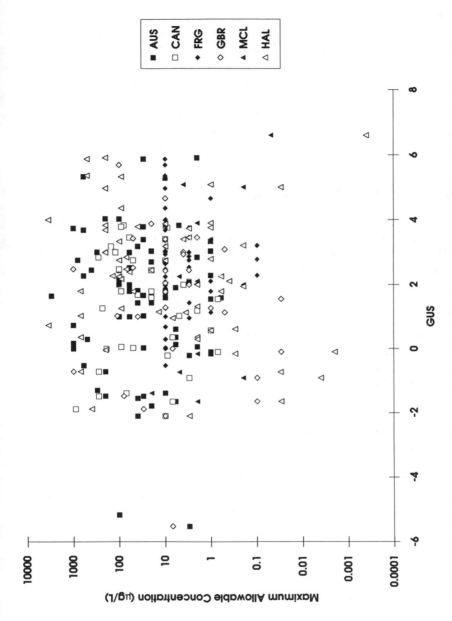

FIGURE 5-1. Maximum acceptable concentrations of pesticides in water as a function of groundwater ubiquity score. No pattern or functionality is implied by this graph; it is merely intended to convey pictorially the range of values observed.

poor agreement between countries on what constitutes an acceptable pesticide residue in drinking water. Maximum acceptable concentrations literally vary over eight orders of magnitude across all pesticides and over several orders of magnitude for individual active ingredients (see Figure 5-1 for a plot of these values as a function of the GUS index for each material).

The remainder of this chapter will now focus on the sets of regulations established by the various regulatory bodies. The story begins with the United States, which has the most complex regulatory apparatus.

UNITED STATES REGULATIONS

The EPA is the federal agency responsible for protecting the environment from all causes of pollution and contamination. It enforces the federal laws regarding pesticide use nationally. Currently, the EPA administers five major statutes regarding drinking water:

1. The Federal Insecticide, Fungicide, and Rodenticide Act (FIFRA), under which the EPA regulates the marketing and use of pesticides;
2. The Safe Drinking Water Act (SDWA), under which the EPA sets maximum contaminant levels (MCLs) for contaminants such as nitrates and pesticides that may be present in public drinking water supplies;
3. The Clean Water Act (CWA), under which the EPA is obliged to develop overall drinking water protection strategies;
4. The Resource Conservation and Recovery Act (RCRA), under which the EPA regulates the storage, transportation, treatment and disposal of hazardous wastes by manufacturers and distributors of all chemicals, including pesticides; and
5. The Comprehensive Environmental Response, Compensation and Liability Act (CERCLA), which is usually known as Superfund and finances the government-enforced cleanup of drinking water contamination resulting from chemical spills of accidents at manufacturing plants and storage facilities.

The U.S. Geological Survey (USGS), a division of the Department of Interior, is responsible for providing hydrologic information and appraising the quantity, quality, and use of the nation's water resources.

The U.S. Department of Agriculture (USDA) is responsible for the Water Quality Initiative, a project in which various efforts to protect drinking water from pesticide contamination are underway. The various parts of the USDA that are involved in this effort include:

1. The Soil Conservation Service (SCS), which is the primary technical resource for data concerning soil properties affecting pesticide transport into drinking water;
2. The Cooperative Extension Service (CES), which provides educational programs on planning and installing conservation practices;
3. The Agricultural Stabilization and Conservation Service (ASCS), which administers the Agricultural Conservation Program (ACP), a national cost-sharing program for conservation practices, installed with SCS technical assistance;
4. The Agricultural Research Service (ARS), which conducts research projects on land use and management systems and their effects on pesticide transport into drinking water;
5. The Economic Research Service (ERS), which has established an economic framework to collect data, evaluating the economic sustainability of alternative production systems proposed to benefit drinking water quality; and
6. The Forest Service (FS), which is working on conservation plans to protect water quality on Forest Service lands.

The EPA recently published its Pesticides and Ground Water Strategy (1991a), which describes the policy framework within which the EPA intends to address the risks of groundwater contamination by pesticides. The strategy evolved over many years out of a give-and-take process between the many interested parties, including representatives from federal agencies, state agencies of agriculture, environmental, and health, industry groups, environmentalists, growers, environmental scientists, and Congressional staff.

The first draft of this strategy was issued in February 1988. Public comment on the draft document was extensive, and the finalized strategy incorporates many of the recommendations made by the public during the comment period.

In publishing this strategy, the Agency acknowledged that pesticides were only one of many potential threats to groundwater, with nitrates, salt water intrusion, and exhaustion being among the other important factors. It was also emphasized that there is no major threat to human health posed by pesticides in groundwater at this time and that the major thrust of the strategy was to prevent future contamination by careful vigilance of the chemical products now being used or planned in the future to be used as pesticides.

The strategy is precedent setting in that it establishes for the first time a very close working relationship between the states and the EPA in an effort to place the responsibility for developing management plans for specific "problem" pesticides squarely in the hands of the state. This is accomplished via a three-pronged strategy, as detailed below:

1. An environmental goal;
2. A prevention policy and program; and
3. A response policy and program.

Environmental Goal

Reflecting the overall Agency goal of preventing adverse effects to human health and the environment, the strategy's main goal is to prevent contamination of groundwater resources that present an unreasonable risk of adverse effects to human health and the environment, resulting from the normal, registered use of pesticides.

There are a number of points in this goal that require further clarification. By "groundwater resources," the Agency goes on to state that it means current sources of potable water, reasonably expected sources of such drinking water, as well as groundwater that may comingle with or otherwise affect directly connected surface water bodies and ecosystems.

When determining "reasonable expected sources" of drinking water, such factors as location, intrinsic water quality, population patterns, and the availability of other sources need to be considered.

As pointed out previously in this chapter, the Agency's primary mechanisms for the protection of surface water sources of drinking water, which are typically publicly owned, is the promulgation of MCLs for pesticides and other potentially toxic contaminants of drinking water. In the case of groundwater, which often involves privately owned wells when used as a potable supply, the MCL is invoked by the strategy to serve as a "reference point." When levels of pesticides in drinking water are observed at levels in excess of the "reference point," it would generally be taken as evidence that existing methods for controlling the use of the pesticide are unacceptable.

Where MCLs are not available (only a few have been published, see Appendix 2), EPA Health Advisory Levels or other health-based levels are recommended in the strategy for use as reference points.

Under FIFRA, the observation of pesticides in drinking water at levels above the MCL or other reference point would not necessarily be grounds for immediate regulatory action to cancel or suspend use of the product. The Agency is bound to weigh the benefits of continued use against these risks in order to develop a sound regulatory approach. For instance, it is possible that economic costs to society and/or the potential dangers of replacement pesticides would cause the Agency to permit the occasional occurrence of a toxic pesticide in drinking water. However, the Agency does explicitly state in the strategy that the risks posed by pesticide contamination at levels above the reference point would generally deserve a careful assessment of continued pesticide use in that area.

Prevention Policy and Program

Essentially everyone involved in the manufacture, use, and regulation of pesticides recognizes that the optimal strategy is to prevent their movement into drinking water supplies, as opposed to using them carelessly and cleaning up the problems later on. Several programs have been developed and implemented by the Agency to carry out this stated policy:

1. State groundwater programs;
2. Wellhead protection program;
3. Nonpoint source programs;
4. Sole source aquifer program; and
5. Public water system supervision.

These existing programs are enhanced by the development of a sound regulatory strategy. The strategy described by the Agency involves four steps.

Step 1—Assess Leaching Potential

Using the standard principle that high mobility and high persistence lead to the potential for leaching, the EPA routinely assesses the leaching potential of all pesticides. As part of this process, the EPA has required new environmental fate data from registrants of both old and new products in order to identify those with physical-chemical characteristics indicating a leaching potential. Screening of data received on 800 pesticidal active ingredients by the Agency during the early to mid-1980s indicated that 141 had one or more characteristics of pesticides that had been detected in groundwater in the United States (Behl et al. 1990). The Agency has also required field monitoring studies of varying levels of complexity in order to answer this question more definitively.

The field studies break into four categories of increasing complexity:

1. Field dissipation studies;
2. Small-scale prospective leaching studies;
3. Small-scale retrospective studies; and
4. Large-scale retrospective studies.

The specifics of study design and required data have still not been finalized by the EPA. The following issues are still not settled: triggers for when the various types of studies will be required, standardized site selection procedures, whether the use of tracers (such as bromide ion) are to be required, standardized soil coring procedures, use of suction lysimeters, and placement and sampling of wells. The issues of the design, conduct, and limitations of each of the four types of studies are discussed in the following paragraphs.

Usually, field dissipation studies are conducted using normal agronomic practices, especially for those compounds applied to the soil surface or incorporated into the soil. For foliar-applied materials, however, it may be necessary to conduct side-by-side tests in which the pesticide is applied both to bare soil and in its normal manner. The characterization of pesticide dissipation under field conditions is fraught with technical problems, among them a general confusion of the terms degradation and dissipation (which might include volatilization, runoff, and other modes of off-site movement), the estimation of kinetics, poor sampling methods, and inadequate statistical treatment of data. The number and location of soil cores collected during a single sampling interval is greatly influenced by the inherent variability of soil samples as well as the application. Studies (Thompson et al. 1984) indicate coefficients of variation ranging from 100 to 200 percent. Therefore, usually 15 to 20 soil cores must be collected at each sampling interval (Walker and Brown 1983; Smith, Parrish, and Carsel 1987). If necessary to reduce analytical costs, these cores can be composited in the laboratory to a smaller number of samples. Care should be taken as the experiment proceeds, however, as there is some evidence that the variability increases with time (Walker and Brown 1983). Soil sampling intervals should be tailored to the expected dissipation rates of the agricultural chemical under study. For field studies designed to address groundwater concerns, these intervals may be significantly different from those normally used in field dissipation studies. A typical sampling schedule would be to collect soil samples prior to and immediately after application, at 0.25, 0.5, 1, and 2 months, and at 2-month intervals thereafter until the dissipation pattern of the parent compound and its significant soil degradates has been established. Some of the earlier intervals may be eliminated for pesticides that dissipate more slowly. Soil cores are sampled in 0.15-meter increments down to at least 0.9 meters below the soil surface. The conduct of dissipation studies in fields that are subsequently flooded, such as rice paddies, presents unique challenges (Ross et al. 1989; Nicosia et al. 1991), as does their conduct in forest ecosystems (Thompson et al. 1984; Roy et al. 1989a, 1989b).

The second type of field study is the small-scale prospective study (Jones and Norris 1991), which is generally triggered if residues are found below 0.75 meters in the previously described field dissipation study. Typically, a plot size of 1 to 2 ha is used for these studies. The study objectives determine the soil characteristics sought in the site selection process. If the objective is to measure movement and dissipation under typical agricultural conditions, then a typical soil should be chosen. However, studies must often be conducted in soils that favor residue movement (typically soils with low organic matter and low water holding capacity). The water table depth is also a key factor in site selection. Usually, an area with shallow water tables (less than about 6 meters below the soil surface) is desirable for a study such as this that includes saturated zone monitoring.

Soil sampling in the small-scale prospective study is done in a similar fashion as in the field dissipation study, except that thicker increments to deeper depths are typically used: 0.3-meter increments down to 0.6 meters and 0.6 meters thereafter. These depth increments are significantly larger than the 0.15-meter increments currently required in field dissipation studies. However, unsaturated zone residue plumes of mobile compounds below about 1 to 2 meters are often spread over vertical distances exceeding 1 meter.

The number and location of monitoring wells depends on site characteristics and on whether residues reach groundwater during the study. One approach that has been used successfully in a number of studies is to install an initial grid consisting of five or six well clusters that monitor the upper 3 meters of the saturated zone. A typical well cluster is composed of wells screened at different depths so that the vertical position of any residue plume can be determined. For field studies conducted in areas where the water table is relatively shallow, the clusters have generally consisted of three wells per cluster with 0.3-meter long slotted screens located just below the water table and at 1.5 and 3 meters below the water table at the time of installation. If residues are found in shallow groundwater, then this well network is expanded as needed by adding additional well clusters or deeper wells in existing clusters. The direction of groundwater flow (determined using water table elevation measurements from the initial well clusters) and the pattern of residues in the monitoring wells can be used to select the location and depth of additional wells. Samples should be collected from monitoring wells prior to application and at regular intervals thereafter. Because agricultural chemical residues are applied over a relatively large area while movement of residue plumes in the saturated zone are usually no more than about a meter a day (typically less than 0.2 m/day), water samples normally do not need to be collected in response to recharge events. One common schedule is to sample monthly during the first 6 to 12 months after application and, if necessary, every 2 or 3 months thereafter. If residues are detected in monitoring well samples, sampling is usually continued until the saturated zone dissipation rate is determined or until residues drop detectable levels in all wells for two consecutive samplings.

A controversial issue in the conduct of small-scale prospective studies is whether suction lysimeters will be utilized. Data showing the inadequacy of suction lysimeters to describe real concentration profiles have been reported (Shaffer, Fritton, and Baker 1979). The Agency has requested soil pore water data to be collected because of the higher detection limits generally associated with soil analyses (10 μg/kg) than with water analyses (0.1 μg/L) and because of dilution effects caused by taking larger soil samples with depth and compositing. The Agency feels that soil sampling alone may overestimate the rate at which pesticide residues dissipate in the environment.

The third type of field study, the small-scale retrospective experiment, involves well water sampling in vulnerable regions of known historic pesticide use. Several small-scale retrospective studies have been completed (Roux, Hall, and Ross 1991), but interpretation of the results is difficult because of two main factors:

1. Minimal site characterization is required.
2. Only groundwater is sampled, generally making it impossible to determine whether additional soil residues are likely to enter the groundwater, and there are no good data on pesticide degradation rates.

Nevertheless, the Agency still feels these studies are appropriate when the monitoring sites are selected carefully and the fate properties of the pesticide in surface soils are already well established. When contamination is found in such a study, it may trigger the large-scale retrospective studies.

Large-scale retrospective studies are essentially national surveys of drinking water quality, such as was conducted by Monsanto in the National Alachlor Well Water Study (see Chapter 2). These studies demonstrate what residues are actually present in drinking water on a national basis at the time they are conducted. They do not provide data on mechanisms of contamination, but they provide invaluable information for completing the detailed risk-benefit analysis mandated under FIFRA: do the residues of the pesticide present in drinking water present an unreasonable risk given the benefits of continued use of the material?

Step 2—Assess Effectiveness of National Label
The results of all such testing are generally used to determine whether to register a chemical and to determine appropriate labeling precautions and use directions. Under FIFRA, labeling requirements are legally enforceable, as well as being the primary vehicle for conveying precautionary information to the pesticide user. As it relates specifically to concerns over groundwater contamination, the label directions are intended to minimize the likelihood of either direct (point-source) or indirect (nonpoint-source) contamination. Based on the environmental fate data, the EPA will specify appropriate restrictions on the label, such as maximum rates of application, seasonal timing of applications, and uniform requirements such as well set-backs or anti–back-siphoning requirements.

Step 3—Assign Restricted Use Status

In addition to specifying the legal authority involved in labeling procedures, FIFRA grants the authority to the EPA to classify a pesticide as "Restricted Use"—whereby only trained and certified applicators or persons under their direct supervision can apply the product. Other detailed aspects of application practices can also be specified as part of the conditions of Restricted Use. Up until the late 1980s, the only criterion that could cause a pesticide to be classified as a Restricted Use material was its exceedence of certain toxicity thresholds; however, as a result of the several instances of drinking water contamination, the Agency is now establishing specific criteria under which likely drinking water contaminants would be classified as Restricted Use. Criteria to be used in this classification scheme include chemical characteristics or monitoring data showing detections in drinking water.

Details of the EPA's proposed criteria were given in draft form in 1991 (EPA, 1991b). Two options were proposed. One option would add criteria consisting of either the measured persistence and mobility of a pesticide, or the detection of the pesticide in groundwater in at least three different counties. This was EPA's preferred option in May 1991. The second option would consider the pesticide's persistence and mobility and/or (depending on the product's data base) whether the pesticide had been detected in groundwater in at least three different counties at levels greater than 10 percent of the MCL or lifetime HAL, or in 25 or more wells in 4 or more states.

In proposing these criteria, the EPA indicated a belief that laboratory data on persistence and mobility are useful measures for predicting whether a pesticide has the potential for reaching groundwater. In order to be considered for Restricted Use, the pesticide must meet both the persistence and mobility criteria. For mobility, the proposed criterion is (1) any soil/water partition coefficient, K_D, less than or equal to 5 L/kg, (2) any carbon referenced sorption coefficient, K_{OC}, less than or equal to 500 L/kg, or (3) any detection of the pesticide more than 0.75 meters beneath the soil surface. For persistence, the criterion is (1) any half-life greater than 3 weeks in a laboratory aerobic metabolism or field study or (2) any experimentally determined resistance to hydrolytic or photolytic degradation (in soil or water) under any conditions resulting in degradation of less than 10 percent in a 30-day period. These criteria include for consideration data generated in any study conducted in accordance with the Agency's Pesticide Assessment Guidelines or "other reliable scientific data."

These proposals drew fire from a number of quarters, not the least of which was the USDA, which saw them as potentially damaging to the economic health of American agriculture. The department said that EPA's

$33 million estimated impact of the rule would cover only the cost of record keeping for 10 years (*Food Chemical News* 1991). Other costs involving the use of less effective pesticides and the resultant lower crop yields were ignored. The economics of pesticide use and regulation have been studied in greater detail by researchers at the University of California, Berkeley (Zilberman et al. 1991). Pesticide-use fees were shown to be more efficient than outright pesticide bans as a mechanism to attain environmental goals.

For its part, the pesticide industry has also commented via its national trade organization, the National Agricultural Chemicals Association (NACA) (McCarthy 1991). The NACA comments expressed several concerns, including the importance of preserving the significance of the Restricted Use classification, the inappropriateness of national restrictions for what is generally a local problem, the ineffectiveness of the classification in controlling point sources, a disagreement with the EPA statement that laboratory data alone are reasonable predictors for the potential of groundwater contamination, and an assertion that Restricted Use classification should be assigned based on a weight-of-evidence approach, rather than simple numerical triggers.

The new proposals regarding which physical properties suggest a high leaching potential do represent a technical improvement over the criteria first proposed by the in the early 1980s (Cohen et al. 1984). At that time, the list of physical properties most likely to be associated with leaching included such factors as having a water solubility > 30 mg/L. Recent evidence, including extensive detections of simazine (water solubility of 3 mg/L) and the results of the National Pesticide Survey, point more strongly toward persistence and soil sorption coefficients as the key physical property indicators of water contamination potential.

Final establishment of criteria for assigning Restricted Use status, and a new list of pesticides with such status, was expected in early 1993 as this book went to press.

Step 4—Establish SMPs or Cancel

If the EPA has reasonable assurance that the contamination potential posed by a restricted-use pesticide would not cause "unreasonable adverse effects on the environment," then it would continue the registration of the product with only the restricted use status. However, if such assurance is not available, then the EPA would proceed with the final step of its new strategy, which is that registration could only be maintained in states having approved State Management Plans (SMPs). If no states establish approved SMPs, then the pesticide registration is simply canceled due to the groundwater concerns.

Such cancellation decisions are warranted under FIFRA when it can be shown that the continued use of the pesticide poses risks that outweigh its benefits. According to the EPA, cancellation as a response to the potential for contamination of groundwater would generally be appropriate only when there is "persuasive evidence of serious and widespread risk to groundwater." Such a decision would be the subject of intensive interdepartmental government review in which the risk issues are reviewed by the FIFRA Scientific Advisory Panel (SAP) and the benefits information is evaluated by the Department of Agriculture (USDA).

The risk-benefit assessment process in the case of drinking water contamination requires special care because of the degree of local variability. Unlike dietary or applicator exposure issues, which are generally nationwide in scope, drinking water contamination is likely to result in an aggregate risk (as measured by potentially exposed people), which is far smaller. Yet the local adverse health and economic consequences of the degradation of drinking water quality can be severe.

This unique regionality of the problem of drinking water contamination is what prompted the Agency to establish the newly recommended approach of establishing SMPs for each "problem" pesticide. The differential approach minimizes the complementary risks of overregulation for areas where drinking water contamination is not particularly likely or under-protection of highly vulnerable areas that might result from only a national regulatory approach. SMPs could contain a number of different policies in order to prevent unwanted contamination of drinking water supplies, including moratorium areas in which use of the pesticide is not allowed, well construction requirements in use areas, wellhead protection programs, well set-back distances from areas of pesticide use, application controls (rate, timing, method), monitoring requirements, Best Management Practices (BMPs, see Chapter 6), and certification requirements for users.

Response Policy and Program

In the past, response to an episode of drinking water contamination by pesticides has been conducted on an ad hoc basis. The new strategy seeks to correct this situation by developing a well-thought-out approach to contamination response; however, several aspects of the new response policy have been left vague. The EPA charges the states with the primary role in responding to contamination and well mandate that approved SMPs will contain specific plans for actions to be taken in response to observed contamination. One aspect of the response plan that is particularly sticky is the issue of who pays for long-term remedial action at sites previously contaminated through the approved use of a pesticide.

CANADIAN REGULATIONS

To date, the effort to address pesticide contamination of groundwater in Canada has received considerably less attention than that allocated in the United States (McRae 1991). In 1987, the Pesticides Directorate of Agriculture Canada initiated a strategy to deal with the issue of pesticide in groundwater. The strategy was designed in two phases. The first phase has involved mainly information gathering and the second phase will include the development of cooperative programs with other federal departments to maintain regular monitoring of drinking water, research into measures of reducing the potential for contamination, and possible beneficial regulatory actions. A recently issued document (McRae 1991) summarized the Phase I activities of Agriculture Canada. In this document, the regulatory body has rated 86 pesticides on the basis of their potential to leach to groundwater, volume of use, and overall regulatory concerns (toxicity, etc.). The document also describes data specific to Canada on soil and other hydrogeologic features that favor leaching, cropping practices in the vulnerable areas, and an enumeration of the other factors that should be considered in the selection of sampling locations.

Canada assessed the leaching potential of the 86 pesticides through an elaboration of the GUS screening index described in Chapter 4. The original paper described only three categories of leaching potential, but the Canadians chose to expand this to five as shown in Table 5-1.

The volume of use for each of the 86 pesticides was divided into three categories according to the total volume used in Canada per year as follows: large (> 300,000 kg), medium (50,000–300,000 kg), and small (< 50,000 kg). The overall regulatory concern level for each was assessed according to a somewhat subjective rating scheme, including such factors as the extent of use, age, and completeness of supporting data, exposure to humans and the environment, and any other special concerns.

Using these three criteria, the Canadian government published a ranked list of the 86 pesticides (McRae 1991) in descending order of national priority (see Table 5-2). All three criteria were used to develop the list, although

TABLE 5-1   Canadian Categorization of Pesticides According to GUS Values in "Triple A Format"

Category	Criteria for Category	GUS Index Range
C	High leaching potential	>2.8
B	Transition between high and low leaching potential	1.8–2.8
A	Low leaching potential	1.0–1.8
AA	Very low leaching potential	0.0–1.0
AAA	Non-leachers	<0.0

Source: McRae 1991.

TABLE 5-2   Prioritized List of Pesticides Ranked by Canada for Use in the Design of Water Monitoring Programs

Rank	Pesticide	Leaching Potential
1	Atrazine	C
2	Carbofuran	C
3	MCPA	C
4	Metolachlor	C
5	Picloram	C
6	Bromacil	C
7	Metribuzin	C
8	Simazine	C
9	Dinoseb	C
10	Dalapon	C
11	Tebuthiuron	C
12	Terbacil	C
13	Hexazinone	C
14	Metalaxyl	C
15	Aldicarb	C
16	Methomyl	C
17	Trichlorfon	C
18	Oxydemeton-methyl	C
19	2,4-D	B
20	Dicamba	B
21	1,3-Dichloropropene	B
22	Cyanazine	B
23	2,4-DB	B
24	Diuron	B
25	Lindane	B
26	Linuron	B
27	Dimethoate	B
28	Napropamide	B
29	Methamidophos	B
30	Diphenamid	B
31	Metobromuron	B
32	Ethofumesate	B
33	Triadimefon	B
34	Bensulide	B
35	Monolinuron	B
36	Chlorpropham	B
37	Cycloate	B
38	Ferbam	B
39	Disulfoton	B
40	Captan	A
41	Butylate	A
42	Mancozeb	A
43	Triallate	A
44	Fonofos	A
45	Thiram	A
46	Diazinon	A

TABLE 5-2   *(Continued)*

Rank	Pesticide	Leaching Potential
47	Carbaryl	A
48	Sethoxydim	A
49	Phosmet	A
50	EPTC	A
51	Maneb	A
52	Chlorothalonil	A
53	Azinphos-methyl	A
54	Acephate	A
55	Vernolate	A
56	Benomyl	A
57	Dinocap	A
58	Oxamyl	A
59	Pebulate	A
60	Bromoxynil	AA
61	Trifluralin	AA
62	Propanil	AA
63	Chlorpyrifos	AA
64	Malathion	AA
65	Phosalone	AA
66	Ethaflurlin	AA
67	Fenvalerate	AA
68	Parathion	AA
69	Phorate	AA
70	Diclofop-methyl	AAA
71	Glyphosate	AAA
72	Metiram	AAA
73	Methoxychlor	AAA
74	Chlorthal-dimethyl	AAA
75	Permethrin	AAA
76	Dichloropropane	**[a]
77	Pentachloropenol	**[a]
78	Fenitrothion	**[a]
79	TCA	**[a]
80	Methyl Bromide	**[a]
81	Deltamethrin	**[a]
82	Fensulfothion	**[a]
83	Zineb	**[a]
84	Dichloran	**[a]
85	Dichlorvos	**[a]
86	Captafol	**[a]

Source: McRae 1991.
[a]Insufficient information to allow rating.

leaching potential was given the greatest weight and the overall regulatory concern level was given the least.

The final product of the Phase I effort by Canada was a set of province-scale maps showing the areas deemed vulnerable to groundwater contamination. These maps will be utilized in the Phase II efforts to design and implement efficient groundwater monitoring studies to determine whether any contamination of these groundwaters by pesticides has occurred. The maps are color-coded according to two levels of water table depth—greater than and less than 3 meters below the soil surface. The factors considered by Agriculture Canada in developing these maps build upon those used in the development of the DRASTIC index for the United States: soil texture, soil organic matter, soil structure, soil porosity, field (moisture holding) capacity, soil permeability, hydraulic gradient, hydraulic conductivity, pH, temperature, microbial activity, clay content, rainfall, recharge, topography, depth to groundwater, irrigation, types and amounts of pesticide used, and history of pesticide use.

EUROPEAN COMMUNITY REGULATIONS

An effort to harmonize pesticide registration requirements throughout the European Community was recently consummated with the publication of the "Council Directive of 15 July 1991 Concerning the Placing of Plant Protection Products on the Market (91/414/EEC)," officially published on August 19, 1991. Other efforts are underway to standardize required testing throughout the European Economic Community (Nyholm 1991).

Common standards for health and safety are being established within the European Community so that free circulation of pesticide products between members can take place. Another underlying motivation behind the Directive is an attempt to avoid needless duplication of effort and in particular to avoid unnecessary animal experimentation.

The Directive lists which pesticide active ingredients are approved for use in the European Community. Formulations containing these active ingredients are registered at the national level. Other requirements are intended to improve the intra-community exchange of data while protecting confidentiality.

There is specific language in the Directive giving the more environmentally active countries, such as Germany and The Netherlands, some leeway in restricting the use of European Community–approved pesticides within specific water protection zones that have "serious ecological vulnerability problems."

The most controversial of the European Community requirements is the universal drinking water standard (80/778/EEC), which states that no pesticide will be present at concentrations in excess of 0.1 μg/L and that the total of all pesticides shall be less than 0.5 μg/L. This arbitrary standard was established without regard to toxicity and was based mainly on the perception in 1980 that 0.1 μg/L might be a reasonable detection limit for most pesticides in water. It is now known that this standard is routinely violated in many areas, but no regulatory action against any specific pesticide has yet accompanied this observation. Most countries have gone ahead and established health-based maximum acceptable concentrations for pesticides in drinking water, regardless of 80/778/EEC. These are given in Appendix 2 for Germany and Great Britain.

The Netherlands

Due in part to the particularly critical nature of The Netherlands' reliance upon groundwater as a potable water supply, the Dutch have developed some of the most far-reaching techniques for assessing impacts of pesticides on groundwater quality. In particular, a detailed risk assessment scheme was described for determining whether a newly developed chemical may be used as a pesticide in The Netherlands (Jobsen 1990). The scheme relies heavily upon a leaching model developed specifically for Dutch conditions (Boesten and van der Linden 1990).

The criteria developed by the Dutch include the setting of threshold values for pesticide persistence and mobility that could prevent registration (Nijpels 1989). Measured half-lives or DT_{50}s in excess of 2 months are not allowed, nor are soil/water partition coefficients based on organic matter, K_{OM}, of less than 70 L/kg allowed.

Germany

The German Plant Protection Act requires groundwater to be absolutely protected regardless of any known or suspected health hazard to men or animals (Kordel et al. 1991). In a stated effort to assist pesticide registrants in deciding whether lysimeter studies are necessary for their products and to standardize procedures for the conduct of such studies, the German regulatory authority (BBA) recently published a specific guideline document (Fuhr et al. 1990). The German document lists four physical property criteria that, when exceeded, indicate that computer modeling will be needed to be performed to address the need for a lysimeter study. These are:

1. Water solubility > 30 mg/L;
2. Soilwater partition coefficient based on organic carbon, K_{OC}, < 500 L/kg;

3. Soilwater partition coefficient, K_D, < 10 L/kg; and
4. DT_{50} or the first "half-life" > 21 days.

When a pesticide meets one of these criteria, a leaching model such as SESOIL or PRZM is used to calculate the maximum concentration likely to occur in water leaching out of the unsaturated zone. If the predicted concentration exceeds the EEC directive of 0.1 μg/L, then the lysimeter study must be performed.

The lysimeter study is also triggered whenever:

1. In the absence of aged leaching studies, the laboratory leaching tests indicate that the leachate from the experiment comprises greater than 5-percent "persistent" materials (DT_{50} > 100 days); or
2. If > 2 percent of the applied material leaches in the aged leaching study.

The document then goes on to specify the procedures to be used in the conduct of the lysimeter studies. The lysimeters are specified to be undisturbed soil cores with a depth of 1.0 to 1.3 meters and with a surface area of at least 0.5 m^2 and preferably 1 m^2. A loamy or silty sand is to be used such that the total clay and silt constitute less than 30 percent over the entire profile, with the clay making up less than 10 percent. The organic carbon content of the soil is not to exceed 1.5 percent.

Radiolabeled material is to be used, either formulated or technical, and it is to be applied at the highest label rate and at the least favorable time with regard to leaching, as suggested by computer modeling. Every attempt is made to maintain the lysimeters as close as possible to the normal agricultural situation regarding preparation of soil, timing of planting, plant density, fertilizer application, and other practices.

Duplicate lysimeters are suggested but not required, simply as an insurance against any unforeseen problem with the lysimeter. The test is generally performed for a period of two years.

The focus of the experiment is clearly whether the active ingredient or any of its biologically active metabolites are present in the leachate. Except at the end of the experiment, there is no soil sampling. Thus, the lysimeter tests do not provide detailed information on the rate of soil dissipation of the pesticide, and are limited in their overall utility. Carefully conducted field studies of the small-scale prospective type in the United States provide a good deal more information, but if registration in Germany is desired, then one of these tests will need to be performed.

Other aspects of the German registration process have been described in detail (Schnikel, Nolting, and Lundehn 1986). These include laboratory tests on standard soils and field experiments to determine dissipation and mobility factors under more realistic conditions. The importance of pesticide mobility and pesticide persistence in soil were emphasized in another document (Herzel 1987), in which a scheme was developed for classifying the drinking water contamination potential of pesticides. The persistence criterion used was DT_{50} and mobility in soil was characterized by K_{OC}.

A standard set of mathematical models is applied to dissipation data in Germany (Timme, Frehse, and Laska 1986). The six models used include the regression of the logarithm of the residue, the inverse of the residue, or the inverse of the square root of the residue against either time or the square root of time. Of these six models, the two that most commonly give the best fit to field data are those involving the regression of the logarithm of the pesticide residue against either time or the square root of time. As should be clear from the earlier discussion of kinetic theory, the former of these two is the standard linear first-order kinetic model. Regressing the inverse of the residue and the inverse of the square root of the residue against time correspond to 2nd and 3/2 order kinetics, respectively. In every submission of dissipation data to Germany, all six models must be applied to each set of submitted dissipation data.

Sweden

In 1990, the Chemicals Inspectorate published so-called guillotine criteria concerning toxicity and environmental properties that, when triggered by a chemical, would render it "clearly unacceptable" for registration as a pesticide product in Sweden (Andersson et al. 1990). Emanating from a concern over the contamination of drinking water, the guillotine criteria for persistence are shown in Table 5-3.

TABLE 5-3   Swedish Guillotine Criteria for Persistence in Soil

Temperature[a]	DT_{50}[b]	DT_{90}[c]
25°C	>3 months	>10 months
20°C	>4 months	>13 months
15°C	>6 months	>20 months
10°C	>8 months	>27 months

Source: Andersson et al. 1990.
[a]Temperature at which experiment measuring persistence of the pesticide was conducted.
[b]Time for 50 percent of the extractable active substances in the soil to dissipate.
[c]Time for 90 percent of the extractable active substances in the soil to dissipate.

In the case of pesticide mobility, the guillotine criterion established by Swedish authorities is for any active substance having a $K_{oc} < 50$ L/kg and a half-life exceeding 1 month in soil at 20°C.

The methods used to conduct lysimeter studies in Sweden have been more fully described (Bergstrom 1991). In that article, it is recommended that lysimeter studies include two soil types (sand and clay), as close to normal as possible management practices, weekly water sampling, two watering regimes, and the use of double the normal application rates. A variety of methods for collecting soil lysimeters are discussed, and the use of undisturbed soil columns is advocated by a combination of pressing of a coring device and simultaneous removal of adjacent soil as necessary to insert the collector. The lysimeter area should be large enough so as to minimize wall effects (at least 0.2 meters in diameter seems to be necessary). The question of drainage must be addressed, and in particular the question of whether the water at the bottom of the lysimeter should be held under suction. Most investigators have determined that such effects are negligible when the lysimeter is at least 1 meter in length, which most are in this day.

INDIVIDUAL STATE REGULATIONS IN THE UNITED STATES

California

Registration of pesticides in the state of California is tantamount to obtaining registration in a different country. Several special requirements have been established by this politically active state—only two of which will be mentioned here. One is the establishment of Pesticide Management Zones (PMZs) in areas where pesticides have been detected in groundwater as a result of normal use. A PMZ is a geographical area of approximately 1 square mile that has been found to be sensitive to groundwater contamination by the pesticide. Once so designated by the state, future use of the pesticide within any section declared a PMZ is banned. More recently, the state has proposed making PMZs generic to all chemicals on their "leacher" list.

Another of the major efforts by the state to prevent future contamination of groundwater supplies is the setting of Specific Numerical Values or SNVs (Wilkerson and Kim 1986; Johnson 1988; Johnson 1989). Pesticides exceeding these SNV triggers are placed on special lists for subsequent enhanced scrutiny with respect to their potential for drinking water contamination. The documents describing SNVs have been riddled with the misuse of statistical procedures. The approach is also based on the assumption, now known to be faulty, that water solubility is a good predictor of the potential for groundwater contamination.

Florida

Florida, because of its unique dependence on groundwater, has developed the most sophisticated approaches to managing the use of pesticides while maintaining water quality (Hornsby et al. 1990). The approach described by Hornsby includes the use of physical properties, soil properties, and toxicological information to provide the grower with specific guidance on the selection of a pesticide.

Besides this management technique, the state offers a pair of computer programs, CHEMRANK and CMLS, for evaluating whether significant leaching is likely to occur for the pesticides that the grower has selected (Rao, Hornsby, and Jessup 1985; Nofzinger and Hornsby 1986). They are based on a large data base of statewide soil information and simple modeling techniques similar to those described in Chapter 4.

Iowa

The relatively widespread occurrence of atrazine in the drinking water supplies of Iowa has prompted a special statewide regulation governing the use of the chemical (Iowa Dept. of Agric. 1989). All pesticides containing the active ingredient atrazine were classified as Restricted Use, with sales to and use by only certified applicators. Each dealer selling the product was required to file an annual report listing the total volume in gallons or pounds sold.

Further restrictions on the use of the product included a maximum application rate of no more than 3 LB/A per year in most of the state, with an even lower maximum rate of 1.5 LB/A in several counties and townships with known histories of drinking water contamination. Application of atrazine is not allowed by the rule within 50 feet of sinkholes, wells, or surface water bodies. Mixing and handling operations were further restricted to a distance of at least 100 feet from such areas. Specific guidelines for equipment clean-out were given.

The continuing and ever-evolving development of state regulations such as these is to be expected as SMPs are mandated by the EPA and states continue to act proactively in response to contamination incidents.

References

Andersson, L., S. Gabring, J. Hammar, and B. Melsater. 1990. *Unacceptable Properties of a Pesticide.* The Chemicals Inspectorate, Scientific Documentation and Research, Sweden.

Behl, E., C.A. Eiden, D. Wells, and M.A. Barrett. 1990. Monitoring studies as a tool for estimating ground-water exposure. Presented at Conference of the IUPAC, 5–10 August 1990, in Hamburg, West Germany.

Bergstrom, L. 1991. Leaching potential and decomposition of clopyralid in Swedish soils under field conditions. *Environ. Toxic. & Chem.* 10:563–71.

Boesten, J.J.T.I., and A.M.A. van der Linden. 1991. Modeling the influence of sorption and transformation on pesticide leaching and persistence. *J. Environ. Qual.* 20:425–91.

Bowman, B.T. 1988. Mobility and persistence of metolachlor and aldicarb in field lysimeters. *J. Environ. Qual.* 17:689–94.

Bowman, B.T. 1989a. Mobility and persistence of the herbicides atrazine, metolachlor and terbuthylazine in Plainfield sand determined using field lysimeters. *Environ. Toxic. Chem.* 8:485–91.

Bowman, B.T. 1989b. Techniques for studying the mobility and persistence of pesticides using field lysimeters. Presented at Pesticides Directorate Symposium, Agriculture Canada, 28 June 1989, in Ottawa, Canada.

Bowman, B.T. 1990. Mobility and persistence of alachlor, atrazine and metolachlor in Plainfield sand, and atrazine and isazofos in honeywood silt loam using field lysimeters. *Environ. Toxic. Chem.* 9:453–61.

Bowman, B.T. 1991a. Mobility and dissipation studies of metribuzin, atrazine and their metabolites in Plainfield sand using field lysimeters. *Environ. Toxic. & Chem.* 10:573–79.

Bowman, B.T. 1991b. Use of field lysimeters for comparison of mobility and persistence of granular and EC formulations of the soil insecticide isazofos. *Environ. Toxic. & Chem.* 10:873–79.

BPI. 1991. Sunlight cleans up polluted groundwater in test by Seri. *Ground Water Monitor,* 4 Sept. 1991. p. 175.

Cohen, S.Z., S.M. Creeger, R.F. Carsel, and C.G. Enfield. 1984. Potential pesticide contamination of groundwater from agricultural uses. In *Treatment and Disposal of Pesticide Wastes.* Washington D.C.: ACS. pp. 297–325.

EPA. 1991a. *Pesticides and Ground-Water Strategy.* Office of Pesticides and Toxic Substances, Washington D.C. 21T-1022, October 1991.

EPA. 1991b. *Criteria for Classifying Pesticides for Restricted Use Due to Ground Water Concerns,* 56 Fr 22076 No. 92 05/13/91.

Fermanich, K.J., T.C. Daniel, and B. Lowery. 1991. Microlysimeter soil columns for evaluating pesticide movement through the root zone. *J. Environ. Qual.* 20:189–95.

Food Chemical News, 1991. USDA files objection to ground water restricted use criteria. *Pesticide & Toxic Chemical News,* October 16, 1991. pp. 12–13.

Fredrickson, J.K., H. Bolton, S.A. Bentjen, K.M. McFadden, S.W. Li, and P. Voris. 1990. Evaluation of intact soil-core microcosms for determining potential impacts on nutrient dynamics by genetically engineered microorganisms. *Environ. Toxic. Chem.* 9:551–58.

Fuhr, F., M. Herrmann, A.-W. Klein, C. Schluter, F. Herzel, W. Klein, et al. 1990. *Guidelines for the Testing of Agrochemicals as Part of the Licensing Procedure, Part IV., 4-3, Lysimeter Tests to Establish the Mobility of Agrochemicals in the Subsoil,* Issued by the Department of Agrochemicals and Application Technology of the Federal Biological Research Center, Braunschweig, FDR.

Gillham, R.W., M.J.L. Robin, and C.J. Ptacek. 1990. A device for in situ determination of geochemical transport parameters 1. retardation. *Ground Water* 28:666–72.

Goicolea, M.A., J.F. Arranz, R.L. Barrio, and Z.G. DeBalugea. 1991. Adsorption-leaching study of the herbicides metamitron and chloridazon. *Pestic. Sci.* 32:259–64.

Goodrich, J.A., B.W. Lykins, and R.M. Clark. 1991. Drinking water from agriculturally contaminated groundwater. *J. Environ. Qual.* 20:707–17.

Hallberg, G.R., J.L. Baker, and G.W. Randall. 1988. *Utility of Tile-Line Effluent Studies to Evaluate the Impact of Agricultural Practices on Ground Water.* NWWA Proceedings. pp. 298–326.

Herrchen, M., W. Kordel, W. Klein, and R.T. Hamm. 1990. Lysimeter studies of the experimental insecticide bas 263 I. *J. Environ. Sci. Health* B25:31–53.

Herzel, V.F. 1987. Classification of plant protection agents from the point of view of drinking water protection. *Nachrichtenbl. Deut. Pflanzenschutzd.* 39:97–104.

Hornsby, A.G. et al. 1990. Managing pesticides for crop production and water quality protection. *Florida Grower & Rancher.* March 1990.

Iowa Dept. of Agriculture. 1989. *Restrictions on the distribution and Use of Pesticides Containing the Active Ingredient Atrazine or Any Combination of Active Ingredients Including Atrazine,* Section 21.45.41 (206), Iowa Administrative Code. Des Moines, Iowa.

Jobsen, J.A. 1990. *Risk Assessment Scheme,* 2nd Draft for EPPO/Council of Europe Panel on Environmental Risk Assessment of Plant Protection Products.

Johnson, B. 1988. *Setting Revised Specific Numerical Values.* November 1988. Sacramento: CDFA. 24 pp.

Johnson, B. 1989. *Setting Revised Specific Numerical Values.* October 1979. State of California, CDFA, Sacramento, CA, EH 89-13. 12 pp.

Jones, R.L., and F.A. Norris. 1991. Design of field research and monitoring programs to assess environmental fate. In *Groundwater Residue Sampling Design,* ed. R.G. Nash and A.R. Leslie. ACS Symp. Ser. 465; American Chemical Society, Washington D.C. pp. 165–81.

Kordel, W., M. Herrchen, and W. Klein. 1991. Experimental assessment of pesticide leaching using undisturbed lysimeters. *Pestic. Sci.* 31:337–48.

Kordel, W., M. Herrchen, M. Klein, W. Klein, and R.T. Hamm. 1988. *Lysimeter Experiments and Simulation Models to Evaluate the Potential of Pesticides to Leach into Groundwater.* BCPC-Pests and Diseases-1988.

Kubiak, R., F. Fuhr, W. Mittelstaedt, M. Hansper, and W. Steffens. 1988. Transferability of lysimeter results to actual field situations. *Weed Science* 36:514–18.

Lave, L.B., F.K. Ennever, H.S. Rosenkranz, and G.S. Omenn. 1988. Information value of the rodent bioassay. *Nature* 336:631–33.

Leake, C.R. 1991. Lysimeter studies. *Pestic. Sci.* 31:363–73.

McCarthy, J.F. 1991. *Re: Criteria for Classifying Pesticides for Restricted Use Due to Ground Water Concerns*—Document Identification Number OPP-36172, NACA Memo, 12 July 1991. 20 pp.

McRae, B. 1991. *Backgrounder 91-01: The Characterization and Identification of Potentially Leachable Pesticides and Area Vulnerable to Groundwater Contami-*

nation by Pesticides in Canada. Ag Can, Food Production and Inspection Branch, 21 June 1991.

Miyamato, J., W. Klein, Y. Takimoto, and T.R. Roberts. 1985. Critical evaluation of model ecosystems. *Pure & Appl. Chem.* 57:1523–36.

Nicosia, S., N. Carr, D.A. Gonzales, and M.K. Orr. 1991. Off-field movement and dissipation of soil-incorporated carbofuran from three commercial rice fields. *J. Environ. Qual.* 20:532–39.

Nijpels, E.H.T.M. 1989. *Environmental Criteria Concerning Substances for the Protection of Soil and Water.* The Netherlands Ministry for Housing, Regional Development and the Environment, The Hague. 36 pp.

Nofzinger, D.L., and A.G. Hornsby. 1986. A microcomputer-based management tool for chemical movement in soil. *Applied Agricultural Research* 1:50–56.

Nyholm, N. 1991. The European system of standardized legal tests for assessing the biodegradability of chemicals. *Environ. Toxic. & Chem.* 10:1237–46.

Nyholm, N. 1991. The European system of standardized legal tests for assessing the biodegradability of chemicals. *Environ. Toxic. & Chem.* 10:1237–46.

Rao, P.S.C., A.G. Hornsby, and R.E. Jessup. 1985. Indices for ranking the potential for pesticide contamination of groundwater. In Proceedings of the 44th Annual Meeting of the Soil and Crop Science Society of Florida. Jacksonville: SCSSF. pp. 1–8.

Ross, L.J., S. Powell, J.E. Fleck, and B. Buechler. 1989. Dissipation of bentazon in flooded rice fields. *J. Environ. Qual.* 18:105–109.

Roux, P.H., R.L. Hall, and R.H. Ross. 1991. Small-scale retrospective ground water monitoring study for simazine in different hydrogeological settings. *GWMR,* Summer 1991.

Roy, D.N., S.K. Konar, D.A. Charles, J.C. Feng, R. Prasad, and R.A. Campbell. 1989. Determination of persistence, movement, and degradation of hexazinone in selected Canadian boreal forest soils. *J. Agric. Food Chem.* 37:443–47.

Roy, D.N., S.K. Konar, S. Banerjee, D.A. Charles, D.G. Thompson, and R. Prasad. 1989. Persistence, movement, and degradation of glyphosate in selected Canadian boreal forest soils. *J. Agric. Food Chem.* 37:437–40.

Saffigna, P.G., A.L. Cogle, G. McMahon, and B. Prove. 1984. Evaluation of the usefulness of in situ field soil cores for measuring solute movement in a vertisol. *Rev. Rural Sci.* 5:181–83.

Schinkel, K., H.G. Nolting, and J.R. Lundehn. 1986. *Guidelines for the Official Testing of Plant Protection Products, Part IV, Persistence of Plant Protection Products in the Soil—Degradation, Transformation and Metabolism.* Department for Plant Protection and Techniques of Application of the Federal Biological Research Center, Brunswick.

Shaffer, K.A., D.D. Fritton, and D.E. Baker. 1979. Drainage water sampling in a wet, dual-pore soil system. *J. Environ. Qual.* 8:241–46.

Smith, C.N., R.S. Parrish, and R.F. Carsel. 1987. Estimating sample requirements for field evaluations of pesticide leaching. *Environ. Toxic. Chem.* 6:343–57.

Somich, C.J., P.C. Kearney, M.T. Muldoon, and S. Elsasser. 1988. Enhanced soil degradation of alachlor by treatment with ultraviolet light and ozone. *J. Agric. Food Chem.* 36:1322–26.

Thompson, D.G., B. Staznik, D.D. Fontaine, T. Mackay, G.R. Oliver, and J. Troth. 1991. Fate of triclopyr ester (release®) in a boreal forest stream. *Environ. Toxic. & Chem.* 10:619–32.

Thompson, D.G., Stephenson, G.R., Solomon, K.R., and A.V. Skepasta. 1984. Persistence of (2,4-dichlorophenoxy)acetic acid and 2,(2,4-dichlorophenoxy)propionic acid in agricultural and forest soils of northern and southern Ontario. *J. Agric. Food Chem.* 32:578–81.

Timme, G., H. Frehse, and V. Laska. 1986. Statistical interpretation and graphic representation of the degradational behavior of pesticide residues. II. *Planzenschutz-Nachrichten* 39:187–203.

Walker, A., and P.A. Brown. 1983. Spatial variability in herbicide degradation rates and residues in soils. *Crop Protection* 2:17–25.

Wauchope, R.D., R.G. Williams, and L.R. Marti. 1990. Runoff of sulfometuron-methyl and cyanazine from small plots: effects of formulation and grass cover. *J. Environ. Qual.* 19:119–25.

Wilkerson, M.R., and K.D. Kim. 1986. The pesticide contamination prevention act: Setting Specific Numerical Values. Sacramento: CDFA. 27 pp.

Yu, C.-C., G.M. Booth, D.J. Hansen, and J.R. Larsen. 1975. Fate of alachlor and propachlor in a model ecosystem. *J. Agric. Food Chem.* 23:877–79.

Zilberman, D., A. Schmitz, G. Casterline, E. Lichtenberg, and J.B. Siebert. 1991. The Economics of pesticide use and regulation. *Science* 253:518–23.

6

Improvement of Pesticide Handling

The impact of the regulatory efforts described in Chapter 5 has been significant in the struggle to reduce the occurrence of pesticides in drinking water, but it has been dwarfed by the recent revolution in methods of pesticide handling. Yesterday's carefree methods are being revolutionized through new application technologies and an extensive effort to educate users about proper techniques for handling pesticides before, during, and after application.

BEST MANAGEMENT PRACTICES

The primary vehicle for delivering new information concerning better application and handling techniques is the use of Best Management Practices (BMPs). For the most part, BMPs are simply a formalization of "common sense" practices that any logical person would adopt, while remaining conscious and vigilant of the possible impact of pesticide-related activities on drinking water quality.

Filter Strips

Beginning in 1986, the United States Department of Agriculture began a program in which the use of vegetative filter strips along the sides of streams, rivers, ponds, and lakes was encouraged by paying farmers a dollar amount to take the land out of production. Administered by the USDA Economic Research Service, the Conservation Resource Program (CRP) covers five different environmental provisions. The land categories are irrigated land on highly saline soils, irrigated land in groundwater depletion areas, erodible land in watersheds with high sediment and nutrient pollution problems, buffer strips near streams, and cropped wetlands.

The benefit of such filter strips is many-fold, but their main effect on water quality is to virtually remove all sediment, excess nutrients, and pesticides from runoff water before they can enter the adjacent surface water body. Besides improving drinking water quality through reduced pesticide surface runoff loads, they help stop erosion and provide wildlife habitat (*Conservation Impact* 1991a). Leaving vegetative filter strips around wetlands can reduce the risk of saline soils developing. Saline soils frequently occur when wetlands are cultivated too close to the water's edge.

The establishment of permanent cover when implementing filter strips provides food and cover for a number of wildlife species. Such strips provide attractive areas for nesting upland game birds and waterfowl. These areas also provide important escape cover and travel lanes around the water's edge, where it is often most lacking.

The value of vegetative filter strips can be further increased if legumes, shrubs, and trees are included in grass filter strips. Such plantings provide diverse habitats beneficial not only to game animals, but to nongame wildlife as well. These plantings provide permanent soil-holding capabilities.

The vegetative filter strips can be from 66 to 99 feet wide. The only requirement for participation in the program is that the strip be located alongside a stream or other surface water body. The filter strip is planted with grass, shrubs, and trees. Studies have demonstrated that vegetative strips of at least 30 feet in width can serve to filter sediment, nutrients, and other pollutants from agricultural runoff prior to it entering water bodies. The implementation of this practice on a large scale could result in significant water quality improvement within a given watershed.

In a field investigation of the effectiveness of a grassed waterway in reducing 2,4-D loads in surface runoff (Asmussen et al. 1977), it was found that only 30 percent of the herbicide entering the grassed waterway left the system. The remainder was removed from surface runoff through infiltration of runoff water, sediment loss, and sorption of 2,4-D onto vegetative and organic matter. Similar results were obtained in a study involving 2,4-D, silvex, and picloram (Evans and Duseja 1973), in which concentrations of all three pesticides dropped below the limits of detection a few hundred meters from the treatment area.

For intense storms where the amount of rainfall is similar, pesticide runoff from grassed plots was found to be less than half that which occurred from bare soil (Wauchope, Williams, and Marti 1990). There is evidence that similar reductions are achieved, even when no turf is present on the untreated buffer area. In a field study in Texas (Trichell, Morton, and Merkle 1968), researchers found that the slope of the plot and movement over previously untreated soil significantly reduced discharges to surface waters.

One example of the ways in which government agencies are working to help expand the adoption of BMPs is the North Carolina Cost-Share Program for Non-Point Source Pollution Control (*Conservation Impact* 1988a). Enacted in 1984, the program invited volunteering landowners to be paid 75 percent of the costs associated with installing a BMP on their property. The landowner was responsible for providing the remaining 25 percent of the cost, but this could be supplied "in kind." For example, a landowner planning to plant a grass filter strip along a stream could provide the labor and equipment to cover his or her share. The program has been very successful. According to Jim Cummings, the resource program coordinator, "the acceptance and excitement was unbelievable. We got two to three times as many requests for BMP installation than we expected." By involving local (county) governments in the program, the state was able to focus the effort on BMPs that really mattered to local landowners. North Carolina was recognized by the EPA as the national leader for this type of program.

The economic implications of widespread implementation of BMPs has been investigated using a stochastic programming framework (Milon 1983). Results from the planning model indicate that reliability and multiple effluent constraints significantly increase the cost of nonpoint controls, but the effects vary by control alternative.

In an assessment of possible BMPs for reducing pesticide runoff into surface water (Felsot, Mitchell, and Kenimer 1990), it was found that both no-till and contouring significantly reduced pesticide runoff. Although no conservation tillage system completely eliminated pesticide runoff, losses were most effectively minimized by contoured strip-till and no-till, which controlled both water and sediment runoff. In a field study in Maryland (Isensee et al. 1988), short-term effects included groundwater concentrations of atrazine under no-till plots that were temporarily higher than those under conventionally tilled areas.

In a field study of metolachlor transport in surface runoff (Buttle 1990), it was found that herbicide incorporation and contour plowing led to significant reductions in dissolved and adsorbed concentrations, as well as in total metolachlor runoff loads. Incorporation was associated with an increase in the relative importance of sediment in metolachlor surface transport.

IPM

IPM, Integrated Pest Management, implies the use of techniques other than pesticide sprays for the control of pests—whether they are weeds, fungi, or insects. IPM is one of the strategies that the USDA is advocating and researching in order to reduce the current heavy reliance upon chemical pesticides in the agricultural industry. Typical concepts involved in IPM include novel crop rotations, use of beneficial insects, heterogeneous planting in which crops are mixed in the field, and microbial insecticides.

Novel Crop Rotations

In most advanced countries around the world, farmers have come to adopt standard, rigid crop rotation practices. On the face of it, these rotations seem efficient. Not as much capital is required up front since the same equipment is used year after year. The farmer does not have to learn about different crops. In much of the midwestern United States, any crop other corn, soybeans, wheat, and alfalfa is regarded as "trendy." In Japan, certain fields are allocated to continuous rice and that is where they will stay. In much of Europe, conditions are ideal for cereal production and nothing else is planted.

These practices have bred a heavy reliance upon chemical pesticide products. When soil is repeatedly planted to one or one of two alternate crops, it begins to serve as a perfect breeding ground for all of the pests that frequent those crops. Each year the cycle of life begins for the pest, and each year the farmer obliges by religiously planting the crop the pest desires. To be sure, the farmer is generally able to overcome the disastrous economic effects of the pest through a rigorous chemical program, but it is never economical to apply the amount of chemical necessary to achieve 100-percent control. Thus, there is always a residual quantity of weeds, fungus, or insects at the end of each season to spawn the next generation.

To make matters worse, this strategy toughens the pest by selectively removing those members of the population that are more sensitive to the chemical pesticide. Stories of evolved and almost complete resistance to certain pesticide products abound, including phosphate-resistant insects, triazine-resistant weeds, and weeds resistant to the newer ALS-inhibiting herbicides (such as the sulfonyl ureas and the imazolidinones). The grower can exert some influence over this resistance by rotating his use of pesticide products (for instance, by rotating pyrethroid insecticides with the phosphate-type products). However, the use of the first-choice product is usually suggested by the particular crop, pest, and climatic condition faced by the grower, and the alternate product may not be nearly as effective, even with the resistance problem.

As should be obvious, the only way to break this cycle is to institute crop rotation programs that introduce entirely different populations of so-called pests into the soil. While it may not be realistic, as Michael Dukakis so eloquently suggested, to plant Belgian endive in the Midwest, there are other economically viable alternative crops that are being investigated as part of the USDA research program.

Beneficial Insects
Every pest, with the arguable exception of man himself, has its own predator or effective competitor. This fact, coupled with the knowledge of the identity of the potentially beneficial insect, can be used to control the undesirable pest. This can be accomplished directly, by "inoculating" the field with the beneficial insect, or simply by planting the field to a cover crop that is known to serve as a breeding ground for the desirable pest. Research into such beneficial insects continues to be funded by the USDA, and is one of the hottest leads in an effort to control new insect outbreaks in the problem areas of southern California, where much pesticide-resistance has developed.

Heterogeneous Planting
Even the casual gardener knows the beneficial effects of planting certain flowers together. Chrysanthemums, because of their natural pyrethrins, are excellent items to include in any flower bed. Similarly, nitrogen-fixing legumes such as soybeans and alfalfa can replenish soil fertility when they are included in the field. The farmer usually accomplishes this by rotating the entire field into such crops in alternate years, but the same effect could be obtained by alternating planting strips between the two crops.

Microbial Pesticides
By far the most controversial of the techniques for IPM is the use of microbial pesticides, especially when the microbes have been genetically engineered in order to optimize their beneficial effects. Perhaps the most familiar of the nonengineered products is Bacillus Thuringiensis, or B.T., which exudes a toxin known to be effective in the control of many important insects.

Several different strategies have evolved in the genetic engineering industry to introduce the genes causing the production of the B.T. toxin into the genetic makeup of more suitable organisms, both plant and animal. In certain applications, the genes have been transferred directly to the crop in order to allow the plant itself to exert the insecticidal control of the pest. In another application, the B.T.-toxin gene was transferred to a nematode that frequently populates the root zone of corn plants, theoretically providing insect control without the need for the commonly used soil insecticides.

A recent problem that has confronted users of this approach is the evolution of insect resistance to B.T., a development likely to hamper the introduction and success of these ventures.

Low Input Sustainable Agriculture (LISA)

Low Input Sustainable Agriculture (LISA) is a strategy that has received increased attention in recent years as an alternative to classical corn-wheat-soybean rotations that dominate much of the Midwest. The overall concept is to reduce chemical and other inputs to the land while focusing on "sustainable" output of useful crops. By sustainable, most advocates of LISA tend to stress conservation—of water quality, soil fertility, and the soil itself.

Another connotation of the word "sustainable" is in the economic sense for the grower. The plight of farmers in the United States and elsewhere throughout the world has been much the same in recent years. As technology has become more available and more effectively used by the larger, more heavily capitalized growers, yields and overall production have increased, dropping prices. In the face of plummeting value per ton of crop produced, smaller farmers have been squeezed into the choice of jumping onto the expensive bandwagon of purchasing heavy machinery or else getting out of farming altogether. Pressures are also exerted from lending institutions to adopt the newest practices and apply excessive amounts of fertilizer and pesticide as a safeguard of at least adequate returns. The net result is that farming operations are fewer in number and bigger today, and a philosophy of heavily capitalized operations with intensive chemical application strategies has swept much of the world.

As an alternative approach; LISA proponents suggest that some of the cost-benefit strategies be reexamined in order to assess whether overall benefits to the grower are truly optimized. For instance, by always applying enough "cheap" fertilizer to make use of the additional water that comes in that "1-in-10" year, thereby obtaining a huge yield, the farmer is betting an awful lot of money up front. His long-term economic well-being is better ensured by applying only enough fertilizer for the median year. Not only is the grower's balance sheet improved, but the added benefit of lower potential for nitrate contamination is achieved. Choices such as these confront growers at every turn in the process of producing a crop. When the decisions are made with sustainability in mind, then LISA proponents would argue that both the grower and society benefits. As mentioned in Chapter 2, nitrates are by far the most common contaminant in U.S. well water at levels in excess of the MCL. Fertilizer-supplied nitrogen is essential to the healthy development of most crops in most rotations practiced in the United States, and there is little hope of drastically reducing its use. However, some efforts are now underway to develop new ways of gauging the amount of nitrogen that needs to be applied in order to grow an economically optimized crop. It turns out that, while not only damaging groundwater quality, the use of excess of nitrogen fertilizer is also costing farmers a great deal of money in terms of unrecovered costs in the relationship between fertilizer applied and yield.

NOVEL APPLICATION TECHNOLOGIES

Rather simple improvements in application technology can have major benefits in reducing the potential for drinking water contamination by pesticides. One example is a new compressed air direct injection sprayer developed at the University of Georgia's Coastal Plain Experiment Station (*Conservation Impact,* 1991b). Most conventional sprayers have a single tank that holds both the water and the chemical that the farmer uses to apply the pesticide. In the new sprayer, the water and chemicals are held in separate containers and do not combine until they reach a mixing chamber during spraying.

The immediate benefit of the new system is that there is no leftover diluted spray mixture left in the system. This additional material has often simply been dumped on the field or else necessitated costly additional passes to completely exhaust the system in an environmentally sound manner.

The new sprayer uses only precise amounts of water and chemicals. It is calibrated by adjusting the air pressure in the water and pesticide tanks through built-in metering devices. Once spraying is complete, the sprayer stops. Not only are there economic and environmental benefits with the new sprayer, but the potential for applicator exposure to the pesticides is greatly reduced. The mixing step is entirely eliminated, and the sprayer cleaning step involves only the cleaning of the pesticide line with water from the water tank by simply flipping a switch.

The new system also fits in well with the new trend for pesticides to be supplied in reusable or recyclable containers. The reusable container is placed in the new spraying device, used, and then taken right back off without any intermediate steps required.

Another exciting development in new application technologies is a method being examined at the University of Illinois in which sensors on the spray boom react to changes in surface color or reflectance and immediately adjust the output from the nozzle. This technology holds the promise of greatly reducing the amount of pesticide or fertilizer that needs to be applied to the field, by specifically gearing the application rate to spatially varying needs. A part of this puzzle is the need for highly reliable sensors that can rapidly measure the chemical requirements of the soil. Such a tool for measuring nitrogen needs is being developed by James Welsh of Georgia Tech, in conjunction with the University of Georgia's Agricultural Experiment Station at Athens (*Conservation Impact* 1990a). The sensor detects ammonia, which can be released as a gas when fertilizer reacts with soil. The new system would allow farmers to know instantly how much nitrogen a field might need. It could be part of an electronically controlled system applying variable rates of nitrogen fertilizer to the field.

The sensor utilizes a small rectangular piece of glass coated with a material sensitive to ammonia gas. When a laser light enters the glass, beam interference patterns are produced that can be calibrated based on the concentration of ammonia in the air above the soil, with a sensitivity in the <1 mg/kg range. Tests are still underway to determine the practicality of such a system.

A related device is under development by W. L. Fenton of the NSW Agricultural Research Center in Australia (*Conservation Impact* 1990b). Sensors that register the spectral characteristics specific to living weeds can distinguish between weeds, crop residue, and trash. Each of the spray nozzles is independently activated. Whenever desired, the system can be used for broadcast application by switching from detection to a continuous spray mode. Detection and activation are virtually instantaneous. Ground speeds between 6 to 12 miles per hour and spray line pressures of 20 to 40 PSI have been used successfully.

A weed-detecting spray boom can complement other measures of weed control. Where residual herbicides are applied to fallow ground, a lower rate can be used, because weed infestations can later be sprayed with a non-residual herbicide. This reduces the risk of a residual herbicide finding its way into drinking water and carries the added benefit of diminishing carry-over problems.

Many new technologies for the application of pesticides are under investigation at the Laboratory for Pest Control Application Technology (LPCAT) at the Ohio State University campus in Wooster, Ohio (Hall 1988). Some of the strategies under investigation there include electrostatic spray charging, factors affecting droplet impaction and rebound from leach surfaces, control of droplet size, and minimization of spray drift.

Another technology that may impact future applications of pesticide is remote sensing (Moran and Jackson 1991). High-altitude photography and satellite imagery available from SPOT and LANDSAT can be used to measure several critical features related to pesticide application, including soil type, evapotranspiration, crop health, and type of weed varieties.

Another technology that may not seem nearly as "high tech," but that will clearly benefit the environment, is the ever-expanding use of recyclable pesticide containers. One recyclable container of particular note is a stainless steel system introduced by Du Pont in 1991 (*Conservation Impact* 1991c). Called the SST (for stainless steel transport) tank, it is designed to provide long-term, in-the-field service for both the pesticide product dealer and the farmer. It has some advantages over conventional plastic recyclable containers, including longevity, and in reducing the quantity of plastic introduced into the environment.

Use of the 175-gallon SST just five times during the season by a dealer eliminates the need for 350 2.5-gallon pesticide jugs, 175 corrugated boxes, three to four wooden pallets, and many yards of plastic stretch wrap. Since it is a closed system, the SST also reduces worker exposure to pesticides. It is equipped with a double-diaphragm, seal-less pump and a self-contained digital flow meter. Other features include a 19-inch opening at the top for easy cleaning and a bottom outlet valve that allows the user to empty the tank. The SST has double stacking capability to save space, and each tank has an easy-to-read dipstick.

When smaller containers are recycled, cleaning and rinsing is a necessary step. In addition, federal and state laws require that empty pesticide containers must be pressure rinsed or triple rinsed prior to disposal. Proper rinsing of containers ensures that users comply with the law and protects the environment. In addition, users get full value for all of the product purchased.

Pressure rinsing is the simplest and most effective method to prepare containers for recycling. It requires the use of a special nozzle that directs high-pressure water into the container. The nozzles are pushed through the bottom portion of the containers while they are upended over the spray tank and pressure rinsed for 30 seconds.

Triple rinsing is also acceptable, but is considerably more time consuming and only one-third as effective as pressure rinsing. The container is filled at least 20 percent full with water, the lid is replaced, and the container is shaken to rinse the inner surface. The cap is removed, and the container is allowed to drain into the spray tank for 30 seconds. This entire procedure is repeated a total of three times.

Whichever method is used, pesticide containers should be rinsed immediately after they are emptied. Any storage of the initially emptied container prior to rinsing will reduce the effectiveness of the rinsing process because the material left on the inside may dry or "cake up," preventing proper rinsing. Any containers left with a residue pose a potential threat to drinking water.

CONSERVATION TILLAGE

Runoff of soil, sediment, and nutrients represents a major environmental issue regardless of the impact of any pesticides present in the runoff. Conservation tillage, in which the amount of tillage operations is either reduced or eliminated entirely, represents a direct strategy to minimize soil losses and any negative impacts on off-site resources. Much of the effort to introduce conservation tillage has focused on the areas surrounding the Great Lakes and the Chesapeake Bay, but it is practiced widely elsewhere as well.

The adoption of conservation tillage practices is being driven by four main factors:

1. It saves time and work.
2. It reduces soil erosion.
3. It saves fuel.
4. It reduces costs.

In a recent nationwide survey of growers (*Conservation Tillage News* 1987a), 92 percent of the respondents used or recommended conservation tillage and believe that it is increasing in their area. A few suggested a decline in conservation tillage use, indicating that they had encountered technical problems with equipment or decreased profits. In addition, respondents cited attitudes, habits, and weed control difficulties.

There are some potential conflicts between the ideas and concepts of conservation tillage practices and those of sustainable or regenerative agriculture. In a debate held in Ohio in November 1987, proponents of each pointed out that there is significant common ground, but that the perspective of sustainable or regenerative agriculture is much longer term (*Conservation Impact* 1987b). The focus of LISA is on practices such as crop rotations, recycling of animal waste, and over-seeding legumes, which do not always fit well with conservation tillage practices.

One major advantage of conservation tillage systems comes in time of drought. As demonstrated in the major midwestern drought of 1988, a conservation tillage plot exhibited near normal yields with only 2.6 inches of rain during the growing season; at the same time, a nearby conventional field demonstrated virtually no harvest (*Conservation Impact* 1988b).

Although generally recognized as a "good thing," certain researchers have more recently questioned the potential negative impacts of conservation tillage practices on water quality. The logic behind this concern seems, at least superficially, to be sound. As tillage operations are reduced, there is a greater need for certain pesticide applications, especially the so-called burn-down herbicides to remove competitive weeds at the time of planting. In conventional planting operations, these weeds would be removed and the seed bed otherwise readied for the crop through a deep plowing operation. The potential for such operations to loosen the upper few inches of soil and remove the protective cover offered by weeds and the previous crop's litter is primarily responsible for the greatly increased runoff from newly planted fields.

Besides increasing the needs for additional pesticide applications, conservation tillage practices that include the complete removal of all tillage operations have another potentially detrimental effect by allowing for the establishment of relatively permanent and continuous worm holes from the soil surface down as deep as 1 meter.

A research plot in Coshocton, Ohio, demonstrated this phenomenon quite conclusively. The plot was kept in continuous no-till corn for 23 years. Despite the fact that the field had a 9 percent slope, the total runoff over a continuously monitored 7-year period was less than one-half inch. This total came almost exclusively after a single 3-inch downpour in which 2 inches of rain fell within 20 minutes.

Dr. W. M. Edwards, director of the USDA Agricultural Research Service research station that conducted the test, ascribed the remarkable results to the occurrence of earthworms, which, without the disruptive effects of tillage operations, were able to create conduits for water flow that sometimes reached down to the bedrock (at 1 meter). Another lesser factor was said to be the buildup of additional organic matter near the soil surface.

Whether these results have nationwide implications is questionable because less than 5 percent of the nation's cropland is currently under no-till (Dick, Edwards, and Haghiri 1987) and a much smaller fraction of this is under the continuous no-till regime practiced in this research study.

The continuous no-till field has up to 15 nightcrawler holes per square foot of soil surface plus hundreds of holes per cubic foot created by other earthworms, soil insects, and plant roots. The nightcrawlers thrive because they face no tillage disturbance and are surface feeders, attracted by decaying residue and resulting moisture. This heavy earthworm activity is partially responsible for incorporating organic surface residues in the soil, boosting the soil organic matter content to 5 percent in the top inch and 3 percent at 3 inches. At 5 inches below the soil surface, the organic matter content falls back to the 1.5 percent level seen in the conventional tillage plots.

This increase in the soil organic matter content may go a long way toward reducing or even reversing the potential negative impacts of conservation tillage on water quality due to enhanced leaching of pesticide residues. The other beneficial factor is that the increased moisture content of the less-tilled soils should accelerate pesticide degradation rates, which are typically a fairly strong function of soil moisture content.

The net effect of these various positive and negative effects of tillage practices on pesticide transport processes is questionable. In a study typical of the equivocal results that have been obtained, USDA researchers at Beltsville, Maryland, studied a set of side-by-side plots that had been established as either conventional or no-till corn (Helling et al. 1988). The site comprises 20 acres of sandy loam soils with relatively shallow water tables. Applications of the herbicides alachlor, atrazine, and cyanazine were made to the plots annually. Shallow wells were established both above and below a confining clay layer present at approximately 20 feet below the soil surface. Samples from all wells were analyzed regularly for the presence of the three herbicides used. Atrazine was routinely found, and the other herbicides were only rarely seen. No statistically significant differences were observed in the concentrations found in the no-till plots vs. those in the conventional tillage plots. The relative persistence of all three compounds was determined to be the same in the no-till plots as that in conventional plots. Maximum groundwater residues in these studies were 5.9, 3.6, and 1.0 $\mu g/L$ for atrazine, cyanazine, and alachlor, respectively (Isensee et al. 1988).

Another team of researchers has studied the same phenomenon, but in a different location, Ohio. The research, headed by Terry Logan of the Ohio State University Agronomy Department focused on the effects of no-till and fall plowing on pesticide movement in runoff and tile drainage (*Conservation Impact* 1990c). The Ohio scientists compared losses of four widely used herbicides in surface runoff and tile drainage from no-tilled and fall-plowed fields. The fields were in corn/soybean rotations from 1987 to 1989, and the chemicals studied were atrazine, alachlor, metolachlor, and metribuzin. These four herbicides are among those most commonly reported in surface and groundwaters.

Rainfall during the first two years of the study was well below normal, but 1989 was one of the wettest years on record. Tillage practices had no statistically significant effect on the concentrations of the herbicides in either the runoff or tile drain water. The relative losses of all herbicides were much higher in the surface runoff than in the tile drains. The runoff losses were primarily associated with heavy precipitation events just after pesticide application. Thus, the greatest threat to drinking water quality on these relatively heavy soils in Ohio was judged by the research team to be associated with the surface runoff occurring shortly after pesticide application. Tillage practices did not significantly change this fact, nor did they have a significant quantitative impact on the concentrations observed.

In a laboratory study involving the use of intact soil columns from two tillage systems (Clay, Koskinen, and Carlson 1991), it was found that twice as much alachlor leached from surface no-till than from surface conventional tillage columns. The differences in leaching patterns from the surface soil were attributed to the effect of tillage on the soil's physical and chemical properties.

Using pan lysimeters in the field (Hall, Murray, and Hartwig 1989), researchers in Pennsylvania found significantly more movement of atrazine, cyanazine, metolachlor, and simazine through no-till plots than was observed in conventionally tilled plots. The mean areal losses of atrazine and simazine in the no-till soils of these experiments (3.4 percent of applied) were considerably greater than the losses of cyanazine and metolachlor (each 1.6 percent of applied). The tillage effects were again ascribed to the impact of un-collapsed macropores in the no-till plots, allowing greater infiltration.

In a field study of tillage effects on the dissipation of the herbicides imazaquin and imazethapyr (Mills and Witt 1991), it was found that year-to-year variability was greater in the conventionally tilled plots. The variability was sufficient to cause a complete reversal in the effects of tillage system: In 1985, dissipation of the two compounds was faster in the no-till plots, while the reverse was true in 1986. The differences were ascribed to the long period of time that transpired between herbicide application and the first rainfall. In 1986, this critical time period was considerably longer and a combination of photolysis and volatility in the conventional-till plots led to faster dissipation. The presence of the mulch left in the no-till plots was apparently sufficient to buffer these effects and reduce variability in dissipation patterns.

Micro-lysimeter soil columns were used to compare the persistence and mobility of surface-applied pesticides in tilled and no-tilled plots (Fermanich and Daniel 1991). In direct contrast to the results obtained by others, more leaching of carbofuran was observed from the conventionally tilled soil than in the no-till soil columns. The authors hypothesized that a combination of metabolite spectrum changes and enhanced microbial activity in the no-till soil led to less production of a mobile soil metabolite of carbofuran than had leached through the conventional-soil column.

In a computer modeling study attempting to predict the potential impact of conservation tillage practices on drinking water quality (Donigian and Carsel 1987), it was predicted that uniform adoption of reduced tillage practices would reduce the surface water concentrations of today's pesticides by at least 50 percent. Since the study focused on the Lake Erie basin in Ohio, the predicted groundwater concentrations of the simulated pesticides were all very low and the effects of tillage were therefore of negligible importance.

EDUCATION OF USERS

As pointed out in Chapter 3, a high fraction of the incidents of drinking water contamination by pesticides have been caused by carelessness on the part of the user. This is not in any way meant to blame today's user for the problems of pesticides in drinking water today, but there is no doubt that several careless practices of the past contributed to the situation we are faced with today. Many of these past practices were utilized in the name of expedience as opposed to any concern about water quality.

In order to remedy this situation, the pesticide manufacturing industry has undertaken a large effort to educate users about those practices that have been most threatening to water quality. The form that this education has taken has been leaflets, newsletters, videos, and direct training sessions. Most of the information exchanged via these interactions is just plain common sense, but in several cases it needs to be said. The help of the users is needed to ensure that these economically critical tools can continue to be used.

The most common topics covered in these training sessions are the various steps associated with storing, mixing, loading, and applying pesticides, and finally with the cleaning of spray equipment. At many farms, the water used in these procedures is coming from the drinking water well, and convenience (hose length) often dictates that the operations will be performed near the well. Unfortunately, this greatly enhances the probability for contamination to occur.

One significant effort to educate dealers of pesticides was developed by Monsanto beginning in 1989 (*Conservation Impact,* 1989). The company developed and released a comprehensive video training package to help train dealers in environmental safety. The package is called DETAIL (Dealer Environmental Training And Information Library), and it was first released in July 1989. The stated purpose is to assist dealers in meeting the ever-tightening environmental standards and regulations. Topics in the first phase of the library include "Ground Water and Agricultural Chemicals," "Environmental Site Assessment," "Safe Chemical Handling," and "Contingency Planning."

Each training module includes a 17-minute videotape as well as written training and support materials. They address questions commonly asked by chemical dealers, their employees, customers, and communities. Protecting the environment through good stewardship is a theme that is carried throughout the series.

Another significant effort to educate users and dealers has been developed by the pesticide manufacturing industry under the name Alliance for a Clean Rural Environment (ACRE). The ACRE education program contains a number of brochures discussing the many benefits of BMPs and defining many of the processes, such as soil sorption and dissipation, that affect the behavior of pesticides in the environment. By continuing to educate handlers and users of pesticides about water contamination issues and developing new methods for their application and use, it is certain that the incidence of pesticide occurrence in drinking water can be further minimized.

References

Asmussen, L.E., A.W. White, E.W. Hauser, and J.M. Sheridan. 1977. Reduction of 2,4-D load in surface runoff down a grassed waterway. *J. Environ. Qual.* 6:159–62.

Buttle, J.M. 1990. Metolachlor transport in surface runoff. *J. Environ. Qual.* 19:531–38.

Clay, S.A., W.C. Koskinen, and P. Carlson. 1991. Alachlor movement through intact soil columns taken from two tillage systems. *Weed Technology* 5:485–89.

Conservation Impact. 1987a. Scientists research pesticide leaching. Lafayette, Ind.: CTIC. October 1987. p. 1.

Conservation Impact. 1987b. Conservation tillage vs. regenerative agriculture: is there a debate? Lafayette, Ind.: CTIC. December 1987. p. 1.

Conservation Impact. 1988a. North Carolina finds success with local administration in cost-share program. Lafayette, Ind.: CTIC. January 1988. pp. 2–3.

Conservation Impact. 1988b. Drought—a good reason to stick with no-till. Lafayette, Ind.: CTIC. October 1988. pp. 1–2.

Conservation Impact. 1989. Monsanto develops environmental training videos for dealers. Lafayette, Ind.: CTIC. June 1989. p. 1.

Conservation Impact. 1990a. Low-cost sensors could cut fertilizer use, costs. Lafayette, Ind.: CTIC. May 1990. p. 6.

Conservation Impact. 1990b. Spraying only where needed: weed sensors activate sprayers on-the-go. Lafayette, Ind.: CTIC. September 1990. p. 3.

Conservation Impact. 1990c. Studying pesticide movement in soils: no-till vs. fall plowing. Lafayette, Ind.: CTIC. January 1990. p. 3.

Conservation Impact. 1991a. Vegetative filter strips provide variety of benefits. Lafayette, Ind.: CTIC. September 1991. p. 5.

Conservation Impact. 1991b. New pesticide sprayer reduces chemical wastes. Lafayette, Ind.: CTIC. June 1991. p. 4.

Conservation Impact. 1991c. Du Pont makes stainless steel mini-bulk system. Lafayette, Ind.: Lafayette. June 1991. p. 6.

Conservation Tillage News. 1986. Virtually no run-off occurs on 23-year no-till research plot planted in continuous corn. Fort Wayne, Ind.: CTIC. November 1986. p. 1.

Conservation Tillage News. 1987. CTIC nationwide survey consensus says conservation tillage works, is economically sound agricultural practice. Fort Wayne, Ind.: CTIC. February 1987. p. 1.

Dick, W.A., W.M. Edwards, and F. Haghiri. 1987. *Water Movement Through Soil to Which No-Tillage Cropping Practices Have Been Continuously Applied.* pp. 243–52

Donigian, A.S., and R.F. Carsel. 1987. Modeling the impact of conservation tillage practices on pesticide concentrations in ground and surface waters. *Environ. Toxic. Chem.* 6:241–50.

Ehteshami, M., R.C. Peralta, H. Eisele, H. Deer, and T. Tindall. 1991. Assessing pesticide contamination to ground water: a rapid approach. *Ground Water* November–December:862–68.

Evans, J.O. and D.R. Duseja, 1973. Herbicide Contamination of Surface Runoff Waters, Report to the U.S. EPA, EPA-R2-73-266, (Washington DC:US EPA) 99 pp.

Felsot, A.S., J.K. Mitchell, and A.L. Kenimer. 1990. Assessment of management practices for reducing pesticide runoff from sloping cropland in Illinois. *J. Environ. Qual.* 19:539–45.

Fermanich, K.J., and T.C. Daniel. 1991. Pesticide mobility and persistence in microlysimeter soil columns from a tilled and no-tilled plot. *J. Environ. Qual.* 20:195–202.

Hall, F.R. 1988. The laboratory for pest control application technology. In Proceedings of the AABGA Midwest Regional Meeting, 3–5 August 1988, Wooster, Ohio. pp. 88–95.

Hall, J.K., M.R. Murray, and N.L. Hartwig. 1989. Herbicide leaching and distribution in tilled and untilled soil. *J. Environ. Qual.* 18:439–45.

Helling, C.S., W. Zhuang, T.J. Gish, C.B. Coffman, A.R. Isensee, P.C. Kearney, 1988. Persistence and leaching of atrazine, alachlor, and cyanazine under no-tillage practices. *Chemosphere* 17:175–87.

Hileman, B. 1990. Alternative agriculture. *Chem. Eng. News* March 5:26–40.

Isensee, A.R., C.S. Helling, T.J. Gish, P.C. Kearney, C.J. Coffman, and W. Zhuang. 1988. Groundwater residues of atrazine, alachlor and cyanazine under no-tillage practices. *Chemosphere* 17:165–74.

Martin, M.A., M.M. Schreiber, J.R. Riepe, and J.R. Bahr. 1991. The economics of alternative tillage systems, crop rotations, and herbicide use on three representative east-central corn belt farms. *Weed Science* 39:299–307.

Mills, J.A., and W.W. Witt. 1991. Dissipation of imazaquin and imazethapyr under conventional and no-tillage soybean. *Weed Technology* 5:586–91.

Milon, J.W. 1987. Optimizing nonpoint source controls in water quality regulation. *Water Res. Bull.* 23:387–96.

Moran, M.S., and R.D. Jackson. 1991. Assessing the spatial distribution of evapotranspiration using remotely sensed inputs. *J. Environ. Qual.* 20:725–37.

Trichell, D.W., H.L. Morton, and M.G. Merkle. 1968. Loss of herbicides in runoff water. *Weed Science* 30:447–49.

Wauchope, R.D., R.G. Williams, and L.R. Marti. 1990. Runoff of sulfometuron-methyl and cyanazine from small plots: effects of formulation and grass cover. *J. Environ. Qual.* 19:119–25.

7

Developing "Environmentally Friendly" Pesticides

Several factors will conspire to make more environmentally friendly pesticides an imperative in the future. Current estimates indicate that world population will reach 10 billion by 2030 (Schneiderman and Carpenter 1990). Besides this doubling of the population, the newly industrialized countries will be able to afford more and better food, more than doubling the food requirement. In order to meet these needs and the increasing demands for a cleaner environment, it will be critical for society to develop and utilize a number of technological improvements in the production of food. Most researchers now believe that what will be needed is sustainable agriculture, which was defined by the American Society of Agronomy (Lockeretz 1988) as "one that over the long term enhances environmental quality and the resource base on which agriculture depends, provides for basic human food and fiber needs, is economically viable, and enhances the quality of life for the farmer and for society as a whole."

As society becomes even more environmentally sensitive, many currently registered products will be withdrawn from the marketplace. Despite the larger number of people and larger capacity to consume food, there will be fewer people engaged in primary agriculture due to a long-term trend toward mechanization of agriculture. More farmland will be brought under cultivation through the advent of dams, irrigation, and new roads, but this increase is likely to be only incremental. Far in the future, there may be an increase in the amount of indoor farming through the use of cheap lighting and hydroponics, but most food products will be grown outdoors for at least the next century. Under such exposed conditions, there will be a continuing need for many of the pest control products that are now required. Besides meeting food needs alone, it is conceivable that agri-

183

culture will be increasingly relied upon to provide the feedstocks and raw materials for the chemical industries, after today's conventional petroleum-based supplies are exhausted.

Biotechnology has been advocated as one key to the accomplishment of these future technological achievements, because it will allow farmers to reduce their use of pesticides to combat weeds, insects, and disease. Unfortunately, it does not appear as though significant inroads will be made by the biotechnology industry until the second or third decade of the 21st century, because of technical hurdles, economics, and public resistance to the widespread application of transgenic plants, animals, and microbes. However, during this transition period between the 20th and 21st centuries, it is certain that many of today's pesticide products will no longer be used, due mainly to the environmental issues plaguing certain of them. As was seen in Chapter 1, some of these materials have already left the arsenal of tools available to the farmer (DBCP and EDB), and the others mentioned in Chapters 1 and 2 may not be far behind. Environmentally friendly pesticides with low use rates, low toxicity, and less potential for water contamination will accompany the introduction of biotechnology-based products in order to replace the older materials now known to have certain environmental problems associated with their use.

As discussed extensively in Chapter 5, the regulatory community has attempted to force the hand of the pesticide manufacturing industry toward the use of products with more desirable environmental properties. In Europe, the European Community has mandated that no pesticide shall occur at concentrations in excess of 0.1 μg/L in any drinking water supply. In California, the "Big Green" initiative was only narrowly defeated in November 1990, with one of its provisions calling for the outright ban of any pesticide known to cause cancer or birth defects at any level in any laboratory animal. The regulatory problems that have confronted pesticides have been very expensive in terms of lost chemical sales, lower crop yields, and man-hours that could have been spent bringing new, more profitable products to the marketplace. Such regulatory problems are not confined to synthetic organic chemicals. Similar difficulties confront other new products, such as genetically engineered microbial pesticides, pesticide-resistant crops, and BST (a genetically engineered hormone for enhanced milk production).

This new era may also be viewed as a grand opportunity. The public's demands for sound environmental practices will undoubtedly push several current agricultural chemicals out of the marketplace. To take advantage of these changes, it is imperative that the agricultural industry discover and develop products with environmental properties acceptable to the world community. The industry has undertaken several new initiatives to ensure that the product pipeline will be filled with such environmentally friendly products. After giving a more specific definition of what constitutes an "environmentally friendly" pesticide, two aspects of this industry effort will be described: the development of novel formulations to lower contamination potential and the optimization of physical properties to improve environmental behavior.

DEFINING "ENVIRONMENTAL FRIENDLINESS"

This vague term must be defined. In short, an environmentally friendly product is one that:

1. Degrades rapidly and completely after its job is done;
2. Does not leave residues in food or water;
3. Does not move from the target location;
4. Is compatible with manufacturing, formulations, packaging, and application methods that minimize waste and exposure; and
5. Does not perturb the treated ecosystem in any undesirable way.

Each of these desirable features is described in further detail below.

Rapid Degradation

All pesticides must remain active for a finite time in order to have utility against the target pest. However, all pests (weeds, fungi, and insects) pose a threat to the crop only for a certain time, and it would be most desirable for the pesticide to be present only for the time period during which the pest constitutes a significant economic threat. When an active series of chemical analogues has been discovered, metabolic "handles" can be placed within the chemical structure in order to make them more palatable to soil microbes. Certain atoms and functionalities, such as extensive halogenation, generally retard degradation in soil and water, and these types of substitution should be minimized to the extent possible, while still maintaining activity against the target pest.

Lack of Residues in Food and Water

Besides exhibiting rapid degradation after its job is done, a desirable property of the pesticide is to show no detectable residues in either food or water as a result of its normal use. The properties of pesticides that lead to drinking water contamination have already been explored in-depth, but their propensity to occur in food has not yet been addressed. For most herbicides of low toxicity, the residues present in food are very unlikely to pose a significant risk. However, most insecticides have at least some toxicity to humans and livestock, and it is important to determine whether toxic levels will occur in such matrices as a result of either normal or unintentionally high application rates.

In order to answer this question, pesticide manufacturers conduct extensive field studies at labeled and exaggerated rates of application to measure concentrations in the portions of treated crops used for food by animals and humans. Residues in the processed forms of these commodities (for instance peanut butter from peanuts) are also measured. In general, food residues are inevitable when a chemical is applied to a crop. However, low use rate, short persistence, extensive plant metabolism, and/or greater time between application and harvest may lead to undetectable residues in the edible portions of the plant. All of these tactics, as well as changes in agricultural practice, represent ways to minimize pesticide residues in food.

Lack of Movement from Target Location

By staying only at the target location, the identity of which is determined by the target pest, the pesticide is the most efficient it can be and severely reduces its potential for causing undesirable effects. Besides enhancing efficiencies, such pesticide behavior reduces the potential for drinking water contamination and other undesirable off-site effects. As will be seen later in this chapter, low mobility is accomplished through strong binding to soil, often accomplished by making the molecule more lipophilic. Having high lipophilicity confers lower soil mobility, but it is not a panacea. Very hydrophobic agricultural chemicals, most notoriously DDT, have been found to accumulate in the food chain. The potential for this phenomenon to occur is usually quantified by the bioconcentration factor (BCF). The BCF is highest for those chemicals that are very resistant to biochemical attack and possess high lipophilicity, generally measured by the base-10 logarithm of the octanol/water partition coefficient, $logP$ (Isnard and Lambert 1988).

Besides runoff or leaching, another possible route of off-site movement is through volatilization. Depending on the mode of action of the compound, such movement can have highly undesirable consequences, such as the bleaching of nearby (and not-so-nearby) lawns, caused by clomazone in recent years. Extensive volatilization may raise a host of other issues, including contamination of fog and rainwater, or even global climate change for high-sales-volume halogenated compounds. This general subject represents a potential high-visibility environmental topic for the 1990s.

Compatibility with New Delivery Methods

The new farming and application technologies described in Chapter 6 are generally more compatible with certain classes of pesticides, primarily those that are foliarly applied after the crop has emerged. Besides avoiding broadcast applications directly to soil as much as possible, the delivery methods involve novel formulations and systems for delivering pesticides from the manufacturer, to the dealer, and finally to the user. Pesticides compatible with this new philosophy of active ingredient delivery should be advanced over the older products that made extensive use of small, wasteful containers and delivery methods onto bare soil, maximizing the potential for contaminating drinking water.

No Undesirable Perturbation of Ecosystem

Besides the potential for drinking water contamination, there are other potential environmental risks posed by pesticides. Most of these are minimized by ensuring that the nontarget toxicity of the chemical is as low as possible. Everyone is likely to be familiar with the alleged ability of DDT to thin the egg shells of game birds and thereby accelerate their decline. In addition to such indirect damages, certain pesticides pose a direct and imminent hazard to various nontarget species. Bird kills with pesticides formulated as granular products have been documented for a variety of materials and have led to recent regulatory action against one active ingredient in particular—carbofuran.

Other undesirable effects on the environment include fish kills. A series of fish kills in southern Louisiana during the summer of 1991 caused a statewide ban on the use of a commercially important insecticide, azinphosmethyl, used to treat their vast fields of sugar cane (*Food Chemical News* 1991). By late summer, a total of 750,000 fish had died: striped mullet, largemouth bass, yellow bass, freshwater drum, spotted gar, ladyfish, shad, buffalo fish, mosquito fish, carp, southern flounder, blue catfish, and others. Besides the fish, there were allegations that aerial applications of the pesticide had killed "a number of birds, turtles, and alligators at 13 sites" in the state.

While not truly an environmental fate issue, lack of nontarget toxicity is a property that may ameliorate other less favorable environmental traits. In order for risk from a chemical to occur, both exposure and toxicity must be present. The dual importance of the two terms in this equation is often lost on the public, who may not tolerate either, even when the other factor is absent.

In completing this section defining environmental friendliness, it is worth noting one pesticide that is almost certainly environmentally friendly: glyphosate (Rueppel et al. 1977). The compound is rapidly mineralized in soil, shows no potential for runoff or leaching due to a high degree of soil binding, has minimal effects on soil microflora, and is virtually nontoxic to all animal life. The relatively strong binding of the compound to soil appears to be through the phosphonic acid moiety of the molecule because phosphate competes effectively for sorption sites. Since it is virtually inactivated once it strikes soil, glyphosate must generally be applied directly to foliage, but this mechanism fits in perfectly with the new, directed application techniques that are now being developed. As environmental issues continue to guide trends in agriculture, pesticides like glyphosate will play an ever-increasing role.

NOVEL FORMULATIONS TO REDUCE CONTAMINATION POTENTIAL

To some extent, even those pesticides with undesirable properties can be made more acceptable through novel formulations. One topic that has occupied much research attention over the past several years is the development of controlled release formulations to enhance efficacy or improve environmental behavior. These research efforts have resulted in the availability of micro-encapsulated pesticide formulations, which meter the release of active ingredient. One major advantage of such systems is that pesticides that might otherwise degrade too rapidly to be effective against the target pest can be successfully applied as a controlled-release formulation. Such materials with short half-lives have obvious environmental benefits, but would not otherwise be economically viable if applied in a conventional formulation. Controlled-release formulations have been shown to reduce the leaching of alachlor, EPTC, and metolachlor under greenhouse conditions (Koncal, Gorske, and Fretz 1981).

Nitrogen, though not a pesticide, is the agricultural chemical that has been most widely associated with groundwater contamination at levels over its MCL. Slow-release forms of this essential element will certainly become more widely used in the future as one of the means of reducing the occurrence of above-guideline residues of nitrate in drinking water. In the longer term, there will likely be biotechnology-based improvements to soil microbes that effect soil tilth, but for the next several decades there will be a continuing and growing need for better methods of delivering chemical tools to the soil and crop.

OPTIMIZATION AND PREDICTION OF PESTICIDE PHYSICAL PROPERTIES

Besides developing novel and better methods for formulating the pesticide, the pesticide itself can be made to have more optimal physical properties. In order to accomplish this in new pesticide development, two types of information are required:

1. A specific target for the desired physical properties required; and
2. Methods for predicting the effects of chemical structural changes on the physical properties of the pesticide.

Progress has been made on each of these issues, and it is summarized briefly here.

Optimal Range of Physical Properties

The method used here to indicate the desired range of physical properties is to simply plot the physical properties of several pesticides with known drinking water contamination potential on a graph, using different symbols for those known to be contaminants. Two figures are given here, with the first (Figure 7-1) showing pesticides that have been looked for in groundwater,

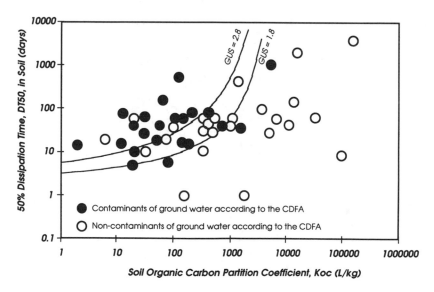

FIGURE 7-1.   Illustration of the influence of pesticide physical properties (mobility and persistence in soil) on the propensity for occurrence at detectable levels in groundwater.

and the second (Figure 7-2) showing pesticides whose occurrence in surface water has been checked. The physical properties forming the horizontal and vertical axes of these two plots are, respectively, the soil/water partition coefficient based on organic carbon, K_{OC}, and the 50-percent disappearance time, DT_{50}. The physical properties of the pesticides are taken from Appendix 1. The status of each pesticide with respect to well water contamination is derived from a list published by the California Department of Food and Agriculture (Johnson 1989), and the surface water contaminants are taken from statewide surveys in Iowa (Iowa DNR 1988) and Ohio (Baker 1983).

As shown in both graphs, those chemicals that possess both high mobility and persistence are more likely to occur in drinking water, whether it is derived from surface or ground sources. The two contours included in each of the figures are defined by values of 2.8 and 1.8 for the GUS index mentioned in Chapter 4. This index, although originally developed to describe specifically the potential for groundwater contamination (Gustafson 1989), also appears to delineate the pesticides likely to be contaminants of surface water (see Figure 7-2). Values of GUS in excess of 2.8 indicate a high potential for occurrence in drinking water, while values below 1.8 indicate a very low potential.

This graphical approach gives a simple and intuitive assessment of a pesticide's threat to drinking water, once its persistence and mobility prop-

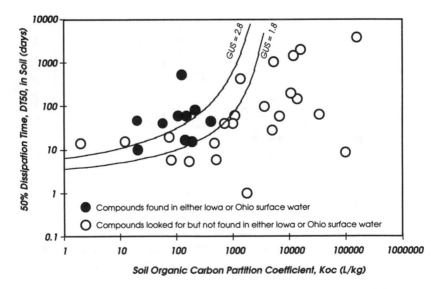

FIGURE 7-2. Illustration of the influence of pesticide physical properties (mobility and persistence in soil) on the propensity for occurrence at detectable levels in surface water.

erties have been obtained. Armed with such a graphical procedure, the developer of new pesticides can attempt to design molecules with properties in the optimal range, which is accomplished by either having very little persistence or low mobility ($K_{OC} > 1000$ L/kg) or both. Some methods for predicting these properties are given below.

Computer Prediction of Physical Properties

Prediction of pesticide mobility is much easier than predicting its persistence. Investigations of the importance of organic matter in the process go back to the 1960s, but in recent years it was probably Karickhoff (in 1984) who showed in a most convincing way the importance of organic matter in the sorption process of pesticides' interactions with soils. He demonstrated that the process is dominated by hydrophobic interactions. Although he states that *a priori* estimation techniques are not yet available, reasonable estimates based upon chemical class and sorbent composition can be made. Relatively accurate predictions of K_{OC} can be made based on water solubility, octanol-water partition coefficient, molecular weight, and reversed-phase HPLC retention time (Kanazawa 1989). Relationships between aqueous solubility and octanol-water partition coefficients have also been described by Mackay, Bobra, and Shiu (1980). One highly nonlinear regression model was recently proposed (Bodor, Gabanyi, and Wong 1989), based on molecular surface, volume, weight, and charge densities on nitrogen and oxygen atoms of the molecule.

Most of the methods for predicting sorptive behavior assume that the pesticide is an uncharged, neutral organic chemical. If ionizable groups are present, then it becomes important to account for the possibility of dissociation and the resulting effects on sorption. In general, molecules carrying a net negative charge will not sorb appreciably to soil; thus, any molecule having a pKa less than 8 will have a very low sorption coefficient. Methods for predicting pKa based on chemical structure have been summarized in a recent textbook (Perrin, Dempsey, and Serjeant 1981).

Certain workers have attempted to predict biodegradation kinetics from conventional structure activity relationships (Desai, Govind, and Tabak 1989). Others have tried the use of molecular connectivity indices and other graph theories to predict both persistence and mobility in soil (Sabljic 1987; Niemei et al. 1987; Hall and Kier 1989; Gerstl and Helling 1987; Sabljic et al. 1990; Gerstl 1990; Karickhoff et al. 1991). In the case of predicting sorption coefficients, these apparently sophisticated approaches reduce to a simple correlation of molecular weight and K_{OC} upon further inspection (Shea 1989).

Besides biodegradation processes, volatility plays a role in determining the dissipation rate of pesticides in the field. Field volatility is a function of vapor pressure, Henry's Law constant, and sorption to soil. Vapor pressures of many nonpolar organic compounds can be estimated by capillary gas chromatography (Bidleman 1984). This approach was used to measure the vapor pressures of several esters of 2,4-D (Hamilton 1980). Henry's Law constants for relatively simple pesticides may be estimated using bond contribution methods similar to those used for estimating partition coefficients (Meylan and Howard 1991).

Together with the simulation modeling tools described in Chapter 4, these methods for predicting physical properties of pesticides present the very real possibility of designing and testing, all on a computer, whether a hypothetical chemical would be likely to contaminate drinking water when used as a pesticide. This promise, when coupled to the rigorous implementation of better application methods, suggests that future contamination of drinking water supplies by pesticides, which is already quite rare, could be entirely eliminated.

References

Baker, D.B. 1983. *Pesticide Concentrations and Loading in Selected Lake Erie Tributaries—1982*, Final Report, U.S. EPA Grant No. R005708-01. EPA, Washington D.C., 61 pp.

Bidleman, T. F. 1990. Estimation of vapor pressures for nonpolar organic compounds by capillary gas chromatography. *Anal. Chem.* 56:2490–96.

Bodor, N., Z. Gabanyi, and C.-K. Wong. 1989. A new method for the estimation of partition coefficient. *J. Am. Chem. Soc.* 111:3783–86.

Borman, S. 1990. New QSAR techniques eyed for environmental assessments. *Chem. & Eng. News* February 19:20–23.

Desai, S.M., R. Govind, and H. H. Tabak. 1990. Development of quantitative structure-activity relationships for predicting biodegradation kinetics. *Environ. Toxic. Chem.* 9:473–77.

Food Chemical News. 1991. Azinphos-methyl linked to kill of half million fish in Louisiana. *Pesticide & Toxic Chemical News*, 14 August 1991. p. 22.

Gerstl, Z., and C. S. Helling. 1987. Evaluation of molecular connectivity as a predictive method for the adsorption of pesticides by soils. *J. Environ. Sci. Health* B22:55–69

Gerstl, Z. 1990. Estimation of organic chemical sorption by soils. *J. Contam. Hydrol.* 6:357–75.

Gustafson, D.I. 1989. Groundwater ubiquity score: a simple method for assessing pesticide leachability. *Environ. Toxic. Chem.* 8:339–57.

Hall, L.H., and L.B. Kier. 1989. Estimation of environmental and toxicological properties: approach and methodology. *Environ. Toxic. & Chem.* 8:19–24.

Hamilton, D.J. 1980. Gas chromatographic measurement of volatility of herbicide esters. *J. Chrom.* 195:75–83.

Iowa Department of Natural Resources. 1988. *Pesticide and Synthetic Organic Compound Survey*. Report to the Iowa General Assembly on the Results of the Water

System Monitoring Required by House File 2303, Iowa DNR, Des Moines. 19 pp.

Isnard, P., and S. Lambert. 1988. Estimating bioconcentration factors from octanol-water partition coefficient and water solubility. *Chemosphere* 17:21–34.

Johnson, B. 1989. *Setting Revised Specific Numerical Values.* October 1979, State of California, CDFA, Sacramento, Calif. Eh 89-13. 12 pp.

Kanazawa, J. 1989. Relationship between the soil Sorption constants for pesticides and their physicochemical properties. *Environ. Toxic. Chem.* 8:477–84.

Karickhoff, S.W. 1984. Organic pollutant sorption in aquatic systems. *J. Hyd. Eng.* 110:707–35.

Karickhoff, S.W., V.K. McDaniel, C. Melton, A.N. Vellino, D.E. Nute, and L.A. Ca. 1991. Predicting chemical reactivity by computer. *Environ. Toxic. & Chem.* 10:1405–16.

Koncal, J.J., S.F. Gorske, and T.A. Fretz. 1981. Leaching of EPTC, alachlor, and metolachlor through a nursery medium as influenced by herbicide formulations. *Hortscience* 16:757–58.

Lockeretz, W. 1988. Sustainable agriculture. *Am. J. Altern. Agric.* 3:174–77.

Mackay, D., A. Bobra, and W. Y. Shiu. 1980. Relationships between aqueous solubility and octanol-water partition coefficients. *Chemosphere* 9:701–11.

Meylan, W.M., and P.H. Howard. 1991. Bond contribution method for estimating Henry's Law constants. *Environ. Toxic. & Chem.* 10:1283–93.

Niemei, G.J., G.D. Vieth, R.R. Regal, and D.D. Vaishnav. 1987. Structural features associated with degradable and persistent chemicals. *Environ. Toxic. Chem.* 6:515–27.

Perrin, D. D., B. Dempsey, and E.P. Serjeant. 1981. *pKa Prediction for Organic Acids and Bases.* New York: Chapman and Hall. 143 pp.

Rueppel, M.L., B.B. Brightwell, J. Schaefer, and J.T. Marvel. 1977. Metabolism and degradation of glyphosate in soil and water. *J. Agric. Food Chem.* 25:517–28.

Sabljic A., H. Gusten, J. Schonherr, and M. Riederer. 1990. Modeling uptake of airborne organic chemicals. 1. plant cuticle/water partitioning and molecular connectivity. *Environ. Sci. Technol.* 24:1321–26.

Sabljic, A. 1987. On the prediction of soil sorption coefficients of organic pollutants from molecular structure: application of molecular topology model. *Environ. Sci. Technol.* 21:358–66.

Schneiderman, H.A., and W.D. Carpenter. 1990. Planetary patriotism: sustainable agriculture for the future. *Environ. Sci. & Technol.* 24:101–23.

Shea, P. J. 1989. Role of humidified organic matter in herbicide adsorption. *Weed Tech.* 3:190–97.

Appendix 1

Properties and Water Quality Criteria for Selected Pesticides

This appendix contains a list of the key physical-chemical properties and health-based maximum allowable concentrations in drinking water for 329 pesticides (Table A.1-1). Pesticides are listed by common name, but there is a look-up list (Table A.1-2) also provided at the end of this appendix by product name. The physical properties given are an average DT_{50} (days) in field soils and an average K_{OC} (L/kg). These average values are generally calculated from a large number of measured values reported in the literature, however a few may be taken from only one source. Space prevents a complete listing of the references for each particular value, but the list of references at the end of this appendix indicates where the values were found.

Following DT_{50} and K_{OC} in the table is the GUS index for that pesticide, calculated on the basis of the given persistence and mobility. As pointed out in the text, values of GUS > 2.8 indicate a high probability for occurring in drinking water, whereas values < 1.8 indicate a very low probability for their occurrence.

The final five column entries indicate the health-based maximum allowable concentrations that have been established by various countries (Australia, Canada, Federal Republic of Germany, United Kingdom, and United States) for the pesticides. If no entry is given for the pesticide under a country heading, then no such concentration had been established by that country at the time this book went to press (late 1992).

Maximum allowable concentrations for the pesticides are not calculated in the same fashion by each country, such that despite the fact that they are generally based on the same toxicology information, there is no reason to

TABLE A-1 Pesticide Properties and Maximum Allowable Concentrations (μg/L) in Drinking Water

Pesticide	DT_{50} (days)	K_{OC} (L/kg)	GUS	AUS	CAN	FRG	GBR	USA[a]
3-CPA	10.0	20.0	2.70					5 (0.6[b])
1,2-dichloropropane	55.0	501.5	2.26			10		(0.2[b])
1,3-dichloropropene	24.0	47.0	3.21			0.1		70 (70)
2,4-D	11.7	47.2	2.48	100	100	10	1,000	
2,4-D (Butoxyethanol ester)	7.0	1,000.0	0.85					
2,4-DB	20.0	20.0	3.51					
2,6-Dichloro-4-nitroaniline	10.0	5,000.0	0.30					(70)
2,4,5-T	22.9	80.1	2.85	2	280			
Accent	20.0	100.0	2.60					
Acephate	3.0	2.0	1.76	20				
Acetochlor	12.8	129.8	2.09					
Acifluorfen	32.0	139.0	2.80					(1[b])
Alachlor	15.4	167.3	2.11	3	[c]	1		2 (0.4[b])
Aldicarb	26.6	23.0	3.76		9	3		(10)
Aldicarb sulfoxide	50.0	20.0	5.09					(10)
Aldoxycarb	20.0	10.0	3.90					(10)
Aldrin	65.0	1,400.0	1.55	1	0.7		0.03	
Alloxydim						10		
Ametryn	48.5	382.7	2.39					(60)
Amidochlor	5.6	247.1	1.20					
Amitraz	20.0	1,000.0	1.30	1				
Amitrole	14.0	100.0	2.29			0.1		
Ancymidol	120.0	120.0	3.99			10		
Anilazine	1.0	3,000.0	0.00					
Arsenic acid	100.0	10,000.0	0.00					
Assert	35.0	35.0	3.79					
Asulam	7.0	40.0	2.03	100	60[d]	10		
Atrazine	81.4	157.3	3.45			3	2	3 (3)

Azinphos-ethyl	40.0					10		
Azinphos-methyl		1,000.0	1.60	10	20			
Barban				300		10		
Benalaxyl						10		
Benazolin						10		
Bendiocarb	7.0	100.0	1.69					
Benefin	43.5	9,000.0	0.07					
Benomyl	220.0	190.0	4.03	200	40			
Bensulfuron-methyl	5.0							
Bensulide	91.3	10,000.0	0.00	400		10		(20)
Bentazon	22.4	149.5	2.46					
Bifenox	7.0	10,000.0	0.00					
Bifenthrin	10.0	100,000.0	−1.00					
Bioresmethrin				60				
Bromacil	265.0	63.2	5.33	600		10		(90)
Bromophos-ethyl				20			110	
Bromoxynil	4.0	190.0	1.03	30	5[d]			
Butachlor	8.0	1,967.9	0.64					
Butylate	51.3	6,163.0	0.36					(700)
Captan	3.0	66.5	1.04					
Carbaryl	14.1	271.3	1.80	60	90			(700)
Carbendazim				200		10	3	
Carbetamide							500	
Carbofenothion				1			0.1	
Carbofuran	34.1	34.2	3.78	30	90	10		40 (40)
Carboxin	4.5	264.0	1.03					(700)
Chloramben	24.0	234.8	2.25		7	10		(100)
Chlordane	1772.8	32,217.3	−1.65	6				2 (0.03[b])
Chlordimeform	60.0	100,000.0	−1.78	20			0.1	
Chlorfenvinphos				10				
Chloridazon	60.0	120.0	3.42			10	50	
Chlorimuron-ethyl	40.0	20.0	4.32			10		

197

TABLE A-1 *(Continued)*

Pesticide	DT_{50} (days)	Koc (L/kg)	GUS	AUS	CAN	FRG	GBR	USA[a]
Chlormequat							10	
Chlorobenzilate	20.0	2,000.0	0.91					
Chloroneb	130.0	1,650.0	1.65					
Chloropicrin	2.0	82.0	0.63					
Chlorothalonil	49.5	1,660.0	1.32					(2[b])
Chloroxuron		4,986.0		30				
Chlorpropham	43.3	1,150.0	1.54					
Chlorpyrifos	37.9	9,149.2	0.06	2	90			
Chlorsulfuron	98.6	455.8	2.67					
Chlorthiamide						10		
Chlortoluron						10	80	
Cinmethylin	30.0	350.0	2.15					
Clomazone	31.3	205.0	2.53					
Clopyralid	30.0	1.4	5.69			10	100	
Cyanazine	38.3	182.7	2.75			10		
Cycloate	30.0	339.4	2.17		10[d]			(10)
Cyfluthrin	60.0	10,000.0	0.00					
Cyhexatin				200				
Cypermethrin	41.5	160,000.0	−1.95					
Cyromazine	90.0	10.0	5.86					200 (200)
Dalapon	30.0	1.0	5.91					
Daminozide	7.0	1.0	3.38					
Dazomet						1		
DBCP	195.0	65.0	5.01					0.2 (0.03[b])
DCPA	100.0	4,333.3	0.73	1000				(3500)
DDT	12419	223,286.7	−5.52	3			7	
Demeton-S-Methyl sulphone	30.0	51.0	3.39	30				
Desmedipham	30.0	2,000.0	1.03					

Diallate	30.0							(0.6)
Diazinon	32.2	654.0	1.79	10	20	1		(200)
Dicamba	29.7	90.7	3.01	300	120	10	4	
Dichlobenil	60.0	194.0	3.04	20		10	40	
Dichlorprop	10.0	1,000.0	1.00			10		
Dichlorvos				20				
Diclofop-methyl	2.0	48,500.0	-0.21	3				
Dicofol	60.0	8,000,000.0	-5.16	100	9			
Dicrotophos	20.0	75.0	2.76					
Dieldrin	917.5	10,800.0	-0.10	1	0.7		0.03	(.002[b])
Diethatyl-ethyl	14.0	1,300.0	1.02				80	
Difenzoquat-methyl sulfate	100.0	54,500.0	-1.47	200				
Diflubenzuron	20.0	9,000.0	0.06					
Diflufenican	60.0	1,000.0	1.77					
Dikegulac						10		
Dimefuron						10		
Dimethipin	10.0	10.0	3.00					
Dimethoate	7.0	12.5	2.45	100	20[d]	10	3	(2100)
Dimethrin								
Dinocap	20.0	630.0	1.56		*c*	3		7 (7)
Dinoseb	30.0	2,236.7	0.96			10		
Dinoterb								
Dioxin		800,000.0						(200)
Diphenamid	30.0	900.0	1.54					20
Dipropetryn	24.0	100,000.0	-1.38					(0.3)
Diquat	5.5	1490.5	0.61	10	70			
Disulfoton	38.0	1,637.9	1.24	6				
Dithiopyr	206.8	422.4	3.18	40	150			(10)
Diuron	10.0	1,000,000.0	-2.00			10		
Dodine								
DPA				500				
DPT+Metabolites					30			

TABLE A-1 *(Continued)*

Pesticide	DT_{50} (days)	K_{OC} (L/kg)	GUS	AUS	CAN	FRG	GBR	USA[a]
EDB	1,470.4	82.3	6.60					0.05 (.0004[b])
Endosulfan	120.0	2,026.7	1.44	40		3		
Endothall	7.0	20.0	2.28	600				100 (140)
Endrin	2,240.0	11,188.0	−0.16	1				2 (0.3)
EPTC	26.2	163.3	2.53	60			50	
Esfenvalerate	35.0	5,300.0	0.43					
Ethalfluralin	40.5	4,000.0	0.64					
Ethephon	10.0	10,000.0	0.00					
Ethidimuron						10		
Ethiofencarb						10		
Ethion	350.0	8,890.0	0.13	6				
Ethofumesate	50.0	37.0	4.13					
Ethoprophos	48.0	103.2	3.34	1		1		
Etridiazole	20.0	10,000.0	0.00					
Etrimfos						10		(0.2[b])
ETU	7.0	50.0	1.94					
Fenac	180.0	20.0	6.09					
Fenamiphos	15.5	202.0	2.02					(2)
Fenarimol	360.0	1,300.0	2.27					
Fenbutatin-oxide	90.0	100,000.0	−1.95					
Fenchlorphos				60				
Fenitrothion				20				
Fenoprop				20				
Fenoxaprop-ethyl	5.0	53,700.0	−0.51					
Fenoxycarb	1.0	1,000.0	0.00					
Fenpropimorph						10		
Fensulfothion				20				
Fenvalerate	35.0	100,000.0	−1.54	40				

200

Ferbam	17.0	300.0	1.87			3		
Flamprop-methyl	20.0	3,000.0	0.68	6		10		
Fluazifop-P-Butyl	21.0	100,000.0	-1.32					
Flucythrinate	20.0	100,000.0	-1.30					
Flumetralin	11.0	80.3	2.18					
Fluometuron	49.8			100				(90)
Fluorodifen	20.0	1,000.0	1.30					
Flurazole	360.0	450.0	3.44					
Fluridone								
Fluroxypyr	45.0	329.0	2.45			10		
Flurtamone	30.0	1,000,000.0	-2.95					
Fluvalinate	140.0	26.0	5.55					
Fomesafen	42.3	1953.2	1.15					(14)
Fonofos	100.0	100,000.0	-2.00					
Formetanate Hydrochloride								
Formothion				100				
Fosamine Ammonium	8.0	150.0	1.65	3,000				
Glufosinate	7.0	100.0	1.69			1		
Glyphosate	40.8	27,525.0	-0.71	200	280[d]		1,000	700 (700)
Haloxyfop					3			
Heptachlor	1,527.3	19,332.5	-0.91	3			0.1	0.2 (.004[b])
Hexachlorobenzene							0.2	1 (0.02[b])
Hexaflurate				60				
Hexazinone	41.9	54.0	3.68	600		10		(200)
Hexythiazox	30.0	6,200.0	0.31					
Hydramethylnon	10.0	1,000,000.0	-2.00					
Imazamethabenz-methyl (M)	35.0	66.0	3.37					
Imazamethabenz-methyl (P)	35.0	35.0	3.79					
Imazapyr	90.0	17.5	5.39					
Imazaquin	81.2	19.0	5.20					
Imazethapyr	90.0	10.0	5.86					
Ioxynil							10	

TABLE A-1 (Continued)

Pesticide	DT_{50} (days)	K_{OC} (L/kg)	GUS	AUS	CAN	FRG	GBR	USA[a]
Iprodione	14.0	1,000.0	1.15					
Isazophos	34.0	100.0	3.06					
Isocarbamide						10		
Isofenphos	14.2							
Isopropalin	100.0	75,000.0	-1.75					
Isoproturon	36.0	129.0	2.94			10	4	
Isoxaben	75.0							
Karbutilat						10		
Lactofen	5.0	100,000.0	-0.70					
Lambda-cyhalothrin	30.0	180,000.0	-1.85					
Lindane	449.8	1,767.1	2.00	100	4	3	3	0.2 (2)
Linuron	65.6	744.5	2.05		c	10	10	
Malathion	1.0	1,638.6	0.00		190		7	
Maldison				100				
Maleic Hydrazide	30.0	20.0	3.99			10		(3500)
Mancozeb	70.0	2,000.0	1.29				10	
Maneb	60.0	1,000.0	1.78				10	
MCPA	13.5	1,000.0	1.13		c	1	0.5	(3.6)
MCPB	14.0	20.0	3.09					0.5
MCPP	21.0	3.0	4.66			1	10	
Mefluidide						10		
Mepiquat	300.0	100,000.0	-2.48					
Mercaptodimethur	20.0	300.0	1.98					
Metalaxyl	21.0	16.0	3.70			10		
Metaldehyde	10.0	240.0	1.62					
Metam-sodium	5.0	20.0	1.89			10		
Metamitron	11.8						40	
Metazachlor						10		

Methabenzthiazuron	89.0	2.0	2.88			10		
Methamidophos	6.0	10,000.0	0.00			3		
Methazole	14.0	400.0	1.85					
Methidathion	21.0	116.0	3.00	60		3		
Methomyl	35.3	80,000.0	−1.88	60				(200)
Methoxychlor	120.0	206.0	2.79		900		30	400 (400)
Methyl bromide	45.0	10.0	2.54			0.1		
Methyl isothiocyanate	7.0		0.36			3		
Methyl parathion	17.5	5,100.3	−2.21	6				
Metiram	20.0	500,000.0			7			(2)
Metobromuron						3		
Metolachlor	35.9	168.3	2.76	800	50[d]	3		(100)
Metoxuron						10		
Metribuzin	48.3	53.0	3.83	5	80	10		(200)
Metsulfuron-methyl	120.0	61.0	4.60					
Mevinphos	3.0	1.0	1.91	6				
Molinate	21.0	50.0	3.04	1				
Monocrotophos	30.0	1.0	5.91	2				
Monuron	166.0	156.6	4.01			10		
MSMA	100.0	10,000.0	0.00					
NAA	10.0	200.0	1.70					
Nabam				30				
Naled	2.0	154.0	0.55					
Napropamide	54.6	316.2	2.61					
Naptalam	14.0	32.0	2.86					
Nitralin				1,000				
Nitrofen	20.4	4,511.0	0.45					
Nitrothalisopropyl								
Norflurazon	90.0	248.0	3.14			10		
Omethoate				0.4				
Oryzalin	41.5	581.3	2.00					
Oxadiazon	72.0	2,360.0	1.16	60				

TABLE A-1 *(Continued)*

Pesticide	DT_{50} (days)	K_{OC} (L/kg)	GUS	AUS	CAN	FRG	GBR	USA[a]
Oxadixyl						10		
Oxamyl	49.5	11.6	4.97			10		200 (200)
Oxycarboxin	20.0	26.0	3.36			10		
Oxydemeton-methyl	10.0	1.0	4.00					
Oxyfluorfen	35.0	52,672.5	−1.11					
Oxythioquinox	20.0	5,000.0	0.39					
Paclobutrazol	182.0	350.0	3.29					
Paraquat	500.0	60,000.0	−2.10	40	10[d]		10	(3)
Parathion	16.7	9,412.5	0.03	30	50	10		
PCNB	21.0	10,000.0	0.00					
Pebulate	16.0	190.0	2.07					
Pendimethalin	273.0	16,300.0	−0.52	600		10		
Pentachlorophenol	48.0	10,587.3	−0.04					200 (200)
Perfluidone				20				
Permethrin	26.5	81,000.0	−1.29	300				
Phenmedipham	36.0	2,740.0	0.88					
Phorate	40.0	1,805.0	1.19		2[d]			
Phosalone	11.3	2,100.0	0.71					
Phosethyl-al	1.0	20.0	0.00					
Phosmet	12.0	612.0	1.31					
Phosphamidon	17.0	1.0	4.92					
Picloram	168.5	23.2	5.87	30	c	10		500 (500)
Piperalin	30.0	5,000.0	0.44					
Piperonyl butoxide				200				
Pirimicarb	275.0	224.0	4.02	100		10		
Pirimiphos-ethyl				1				
Pirimiphos-methyl	10.0	1,000.0	1.00	60		10		
Primisulfuron-methyl	24.7	125.0	2.65					
Prochloraz	120.0	500.0	2.71					
Profenofos	30.0	840.0	1.59	0.6				

Compound								
Promecarb	200.0	356.1	3.33					(100)
Prometon	91.0	575.9	2.43				10	
Prometryn	60.0	990.0	1.79					(50)
Pronamide	6.3	261.1	1.26			10		(90)
Propachlor	30.0	1,000,000.0	-2.95					
Propamocarb	1.0	188.0	0.0	1000				
Propanil	56.0	8,000.0	0.17	1,000				
Propargite	135.0	153.2	3.87	60				(10)
Propazine						10		(100)
Propham		51.0					20	
Propiconazole	20.0	100.0	2.60					
Propoxur	30.0	30.0	3.73	1,000		10		(3)
Pyrazophos				6				
Pyridate	1.0					10		
Quintozene				40				
Quizalafop-ethyl	60.0	100,000.0	-1.78			10		
Sebuthylazine						10		
Sethoxydim	14.3	50.0	2.65			10		
Siduron	120.0	420.0	2.86					
Silvex	23.3	148.4	3.40			10		50 (50)
Simazine	72.1	300.0	1.98		10[d]	10		1 (4)
Sodium dinitro-O-cresylate	20.0	39.6	3.13				10	
Sulfometuron	20.0	550.0	1.44					
Sulprofos	14.0			20				
TCA					[c]			
Tebuthiuron	360.0	80.0	5.36			10		(500)
Temephos	30.0	100,000.0	-1.48	30	280[d]	10		
Terbacil	73.3	45.8	4.36			10		(90)
Terbufos	12.2	3,000.0	0.57		1[d]			(1)
Terbumeton						10		
Terbuthylazine						10		
Thiabendazole	403.0	2,500.0	1.57					
Thiazopyr	145.5	389.0	3.05					
Thidiazuron	10.0	100.0	2.00					

TABLE A-1 *(Continued)*

Pesticide	DT_{50} (days)	K_{OC} (L/kg)	GUS	AUS	CAN	FRG	GBR	USA[a]
Thifensulfuron-methyl	12.0	45.0	2.53					
Thiobencarb	19.0	374.0	1.82	40				
Thiodicarb	7.0	100.0	1.69					
Thiofanox						1		
Thiometon				20				
Thiophanate-methyl	10.0	1,000.0	1.0	100				
Thiram	15.0	383.0	1.67	30				
Toxaphene	9.0	58,408.0	−0.73					5 (0.031[b])
Tralomethrin	27.0	100,000.0	−1.43					
Triadimefon	21.0	273.0	2.07				10	
Triadimenol	28.0							
Triallate	84.5	2,198.3	1.27				1	
Triasulfuron	60.0				230			
Tribufos	10.0	5,000.0	0.30					
Tributyltin chloride		109,250.0						
Trichlorfon	27.0	2.0	5.29	10				
Triclopyr	46.0	234.8	2.71	20		10		
Tridiphane	31.0	5,600.0	0.38					
Trifluralin	94.0	7,101.8	0.29	500		10		(2)
Triforine	21.0	2,000.0	0.92					
Trimethacarb	10.0	200.0	1.70					
Triphenyltin hydroxide	75.0	23,000.0	−0.68		[c]			
Vernolate	12.0	260.9	1.71					
Vinclozolin	20.0	43,000.0	−0.82					
Zineb				30				
Total (ALL PESTICIDES)					100	10		

[a]Values in parentheses are lifetime Health Advisory Levels. Values not in parentheses are Maximum Contaminant Levels.
[b]No lifetime Health Advisory Level is recommended, but this level is considered a "negligible risk" by the EPA and the U.S. National Academy of Science. It is based on a 95-percent upper bound estimate for an excess cancer risk of one in 1 million.
[c]Under review.
[d]Interim Maximum Acceptable Concentration.

TABLE A-2   Synonyms and Product Names for Pesticides Listed in Table A-1

Synonym or trade name	Pesticide(s) referred to by synonym or trade name
2 Plus 2	2,4-D
	MCPP
2,3,7,8-TCDD	Dioxin
2,4,5-TP	Silvex
A-Rest	Ancymidol
Aastar	Flucythrinate
	Phorate
Aatrex	Atrazine
Abate	Temephos
Accelerate	Endothall
Acclaim	Fenoxaprop-ethyl
Accord	Glyphosate
Actellic	Pirimiphos-methyl
Actrilawn	Ioxynil
Alanap-L	Naptalam
Alar	Daminozide
Aldicarb sulfone	Aldoxycarb
Aliette	Phosethyl-al
Ally	Metsulfuron-methyl
Amber	Triasulfuron
Ambush	Permethrin
Amdro	Hydramethylnon
Amiben	Chloramben
Amid Thin W	NAA
Aminotriazole	Amitrole
Amitrol	Amitrole
Amizine	Amitrole
	Simazine
Amizol	Amitrole
Ammo	Cypermethrin
Ansar	MSMA
Antor	Diethatyl-ethyl
Apron	Metalaxyl
Aqua Kleen	2,4-D
Aquathol	Endothall
Aquazine	Simazine
Arelon	Isoproturon
Arena	Alachlor
Argold	Cinmethylin
Arrosolo	Molinate
	Propanil
Arsenal	Imazapyr
Arsonate liquid	MSMA
Asana	Esfenvalerate
Asana XL	Esfenvalerate
Assure	Quizalafop-ethyl
Asulox	Asulam

TABLE A-2   *(Continued)*

Synonym or trade name	Pesticide(s) referred to by synonym or trade name
Atratol	Atrazine
Attrabute	Atrazine
Avadex	Diallate
Avadex BW	Triallate
Avenge	Difenzoquat-methyl sulfate
Azodrin	Monocrotophos
B-Nine	Daminozide
Balan	Benefin
Banvel	Dicamba
Banvel 720	2,4-D
	Dicamba
Basagran	Bentazon
Basudin	Diazinon
Baygon	Propoxur
Bayleton	Triadimefon
Baytan	Triadimenol
Baythroid	Cyfluthrin
Beacon	Primisulfuron-methyl
Belt	Chlordane
Benchmark	Flurtamone
Benlate	Benomyl
Betamix	Desmedipham
	Phenmedipham
Betanex	Desmedipham
Betasan	Bensulide
Bicep	Atrazine
	Metolachlor
Bidrin	Dicrotophos
Bladex	Cyanazine
Blazer	Acifluorfen
Bolero	Thiobencarb
Bolstar	Sulprofos
Botec	2,6-dichloro-4-nitroaniline
	Captan
Botran	2,6-dichloro-4-nitroaniline
Bravo	Chlorothalonil
Bravo C/M	Chlorothalonil
	Maneb
Brodal	Diflufenican
Bromex	Naled
Bronate	Bromoxynil
	MCPA
Bronco	Alachlor
	Glyphosate
Broot	Trimethacarb
Brush-rhap	2,4,5-T
Buctril	Bromoxynil

TABLE A-2   *(Continued)*

Synonym or trade name	Pesticide(s) referred to by synonym or trade name
Buctril+Atrazine	Atrazine
	Bromoxynil
Bueno	MSMA
Bullet	Alachlor
	Atrazine
Butisan	Metazachlor
Butyl 6D	2,4-D (butoxyethanol ester)
Butyrac	2,4-DB
Camphechlor	Toxaphene
Cannon	Alachlor
	Trifluralin
Canopy	Chlorimuron-ethyl
	Metribuzin
Caparol	Prometryn
Capture	Bifenthrin
Carbamate WDG	Ferbam
Carzol	Formetanate Hydrochloride
Casoron	Dichlobenil
Cerone	Ethephon
Chem Hoe	Propham
Chiptox	MCPA
Chlorthal dimethyl	DCPA
Chopper	Imazapyr
Cinch	Cinmethylin
Classic	Chlorimuron-ethyl
Clipper	Paclobutrazol
Cobra	Lactofen
Colonel	Atrazine
Comite	Propargite
Command	Clomazone
Commence	Clomazone
	Trifluralin
Confidence	Alachlor
Conquer	Prometon
Conquest	Atrazine
	Cyanazine
Contain	Imazapyr
Cotoran	Fluometuron
Counter	Terbufos
Crossbow	2,4-D
	Triclopyr
Curacron	Profenofos
Curbetan	Chloridazon
Curtail	2,4-D
	Clopyralid
Cygon	Dimethoate
Cymbush	Cypermethrin

TABLE A-2   *(Continued)*

Synonym or trade name	Pesticide(s) referred to by synonym or trade name
Cyprex	Dodine
Cythion	Malathion
Dacamine	2,4-D
Daconate	MSMA
Daconil	Chlorothalonil
Dacthal	DCPA
Dasanit	Fensulfothion
DCNA	2,6-dichloro-4-nitroaniline
DCP	1,3-dichloropropene
Demon	Cypermethrin
Deploy	Glyphosate
Des-I-Cate	Endothall
Desiccant L-10	Arsenic acid
Devrinol	Napropamide
Di-SYston	Disulfoton
Dibrom	Naled
Dichloropropene	1,3-dichloropropene
Dikar	Dinocap
	Mancozeb
Dimension	Dithiopyr
Dimethazone	Clomazone
Dimilin	Diflubenzuron
Dithane	Mancozeb
	Maneb
DNOC	Sodium dinitro-O-cresylate
Dropp	Thidiazuron
Dual	Metolachlor
Dursban	Chlorpyrifos
Dyfonate	Fonofos
Dylox	Trichlorfon
Dynamite	Dinoseb
Dyrene	Anilazine
Elgetol	Sodium dinitro-O-cresylate
Enide	Diphenamid
Envert	2,4-D
Eptam	EPTC
Eradicane	EPTC
Escort	Metsulfuron-methyl
Esteron	2,4,5-T
Esteron 6E	2,4-D (Butoxyethanol ester)
Ethoprop	Ethoprophos
Ethrel	Ethephon
Ethyl guthion	Azinphos-ethyl
Ethylene thiourea	ETU
Evik	Ametryn
Evital	Norflurazon
Extrazine	Atrazine
	Cyanazine

TABLE A-2   *(Continued)*

Synonym or trade name	Pesticide(s) referred to by synonym or trade name
Fallow Master	Dicamba
	Glyphosate
Far-Go	Triallate
Finesse	Chlorsulfuron
	Metsulfuron-methyl
Florel	Ethephon
Folex	Tribufos
Fore	Mancozeb
Fosetyl-Al	Phosethyl-al
Freedom	Alachlor
Fruitone	3-CPA
Fruitone N	NAA
Fundal	Chlordimeform
Funginex	Triforine
Fungo	Thiophanate-methyl
Furadan	Carbofuran
Furloe	Chlorpropham
Fusilade	Fluazifop-P-Butyl
Galaxy	Acifluorfen
	Bentazon
Galecron	Chlordimeform
Gemini	Chlorimuron-ethyl
	Linuron
Genep EPTC	EPTC
Glean	Chlorsulfuron
Glean C	Chlorsulfuron
	Methabenzthiazuron
Glean TP	Bromoxynil
	Chlorsulfuron
	Ioxynil
Goal	Oxyfluorfen
Golden Leaf Tobacco Spray	Endosulfan
Goltix	Metamitron
Graminon	Isoproturon
Gramoxone	Paraquat
Guardian	Acetochlor
Guthion	Azinphos-methyl
Harness	Acetochlor
Harvade	Dimethipin
Herbicide 273	Endothall
Herbrak	Metamitron
Hoelon	Diclofop-methyl
Honcho	Glyphosate
Horizon	Fenoxaprop-ethyl
Hydrothol	Endothall
Hyvar	Bromacil
Imidan	Phosmet

TABLE A-2   *(Continued)*

Synonym or trade name	Pesticide(s) referred to by synonym or trade name
Isotox	Lindane
Judge	Alachlor
Karathane	Dinocap
Karmex	Diuron
Kelthane	Dicofol
Kerb	Pronamide
Knox Out	Diazinon
Krenite	Fosamine ammonium
Krovar	Bromacil
	Diuron
Kuron	Silvex
Kylar	Daminozide
Laddok	Atrazine
	Bentazon
Landmaster	2,4-D
	Glyphosate
Lannate	Methomyl
Lariat	Alachlor
	Atrazine
Larvadex	Cyromazine
Larvin	Thiodicarb
Lasso	Alachlor
Lasso and Atrazine	Alachlor
	Atrazine
Lasso EC	Alachlor
Lasso II	Alachlor
Lasso Micro-Tech	Alachlor
Lentagran	Pyridate
Lexone	Metribuzin
Limit	Amidochlor
Lo-Vol 6D	2,4-D (butoxyethanol ester)
Logic	Fenoxycarb
Londax	Bensulfuron-methyl
Lontrel	Clopyralid
Lorox	Linuron
Lorox Plus	Chlorimuron-ethyl
	Linuron
Lorsban	Chlorpyrifos
Machete	Butachlor
Manzate	Mancozeb
Marksman	Atrazine
	Dicamba
Mavrik	Fluvalinate
Mavrik Aquaflow	Fluvalinate
Merpan	Captan
Mesurol	Mercaptodimethur
Metaisosystoxsulfon	Demeton-S-methyl sulphone

TABLE A-2   *(Continued)*

Synonym or trade name	Pesticide(s) referred to by synonym or trade name
Metasystox	Demeton-S-methyl sulphone
Metasystox-R	Oxydemeton-methyl
Metham	Metam-sodium
Methiocarb	Mercaptodimethur
Milogard	Propazine
Mitac	Amitraz
Mocap	Ethoprophos
Mocap Plus	Disulfoton
	Ethoprophos
Monitor	Methamidophos
Monurex	Monuron
Morestan	Oxythioquinox
NAA-800	NAA
Nemacur	Fenamiphos
Nemagon	DBCP
Nortron	Ethofumesate
Nudrin	Methomyl
Octachlor	Chlordane
Ole	Chlorothalonil
Omite	Propargite
One Shot	Bromoxynil
	Diclofop-methyl
	MCPA
Option	Fenoxaprop-ethyl
Orbit	Propiconazole
Ordram	Molinate
Ornalin	Vinclozolin
Orthene	Acephate
Orthocide	Captan
Oust	Sulfometuron
Paarlan	Isopropalin
Pay-Off	Flucythrinate
PCA	Chloridazon
PCP	Pentachlorophenol
Penncap-M	Methyl parathion
Penncozeb	Mancozeb
Penoxalin	Pendimethalin
Phosdrin	Mevinphos
Phoskil	Parathion
Phosulphon	Demeton-S-methyl sulphone
Pillarcap	Captan
Pipron	Piperalin
Pix	Mepiquat
Plantvax	Oxycarboxin
Poast	Sethoxydim
Polyram	Metiram
Pondmaster	Glyphosate

TABLE A-2   *(Continued)*

Synonym or trade name	Pesticide(s) referred to by synonym or trade name
Pounce	Permethrin
PP-333	Paclobutrazol
Pramitol	Atrazine
	Prometon
	Simazine
Prefar	Bensulide
Prelude	Linuron
	Metolachlor
	Paraquat
Premerge	Dinoseb
Prep	Ethephon
Preview	Chlorimuron-ethyl
	Metribuzin
Prime	Flumetralin
Prime+	Flumetralin
Princep	Simazine
Probe	Methazole
Prowl	Pendimethalin
Prozine	Atrazine
	Pendimethalin
Pydrin	Fenvalerate
Pyramin	Chloridazon
Pyrazon	Chloridazon
Ramrod	Propachlor
Ramrod and Atrazine	Atrazine
	Propachlor
Ranger	Glyphosate
RE-40885	Flurtamone
Reflex	Fomesafen
Rescue	2,4-DB
	Naptalam
Reward	Vernolate
Rhino	Atrazine
Rhomene	MCPA
Rhonox	MCPA
Ridomil	Mancozeb
	Metalaxyl
Ridomil Bravo	Chlorothalonil
	Mancozeb
	Metalaxyl
Ridomil PC	Metalaxyl
	PCNB
Ro-Neet	Cycloate
Rodeo	Glyphosate
Ronilan	Vinclozolin
Ronstar	Oxadiazon
Roundup	Glyphosate

TABLE A-2   *(Continued)*

Synonym or trade name	Pesticide(s) referred to by synonym or trade name
Rovral	Iprodione
Royal MH	Maleic hydrazide
Royal Slo Gro	Maleic hydrazide
Rubigan	Fenarimol
Rubitox	Phosalone
Ryzelan	Oryzalin
Saddle	Alachlor
Salute	Metribuzin
	Trifluralin
Scepter	Imazaquin
Scout	Tralomethrin
Screen	Flurazole
Sencor	Metribuzin
Sevin	Carbaryl
Sinbar	Terbacil
Slug-Geta	Mercaptodimethur
Solicam	Norflurazon
Sonalan	Ethalfluralin
Sonar	Fluridone
Spectracide	Diazinon
Spike	Tebuthiuron
Spin-Aid	Phenmedipham
Sprout Nip	Chlorpropham
Squadron	Imazaquin
	Pendimethalin
Stall	Alachlor
Stam	Propanil
Stampede CM	MCPA
	Propanil
Stomp	Pendimethalin
Storm	Acifluorfen
	Bentazon
Sulfometuron-methyl	Sulfometuron
Supracide	Methidathion
Surefire	Diuron
	Paraquat
Surflan	Oryzalin
Surpass	Vernolate
Sutan	Butylate
Sutazine	Atrazine
	Butylate
Swat	Phosphamidon
Tackle	Acifluorfen
Tag	Diquat
Talstar	Bifenthrin
Tandem	Tridiphane
TBT	Tributyltin chloride

TABLE A-2   *(Continued)*

Synonym or trade name	Pesticide(s) referred to by synonym or trade name
Team	Benefin
	Trifluralin
Telar	Chlorsulfuron
Telone	1,3-dichloropropene
	Chloropicrin
Telvar	Monuron
Temik	Aldicarb
Tenoran	Chloroxuron
Terraclor	PCNB
Terraclor Super-X	Etridiazole
	PCNB
Terraclor Super-X/Di-Syston	Disulfoton
	Etridiazole
	PCNB
Terraclor Super-X/Thimet	Etridiazole
	PCNB
	Phorate
Terrazole	Etridiazole
Tersan	Benomyl
	Mancozeb
Thimet	Phorate
Thiodan	Endosulfan
Thistrol	MCPB
Tillam	Pebulate
Tiller	2,4-D
	Fenoxaprop-ethyl
	MCPA
Tilt	Propiconazole
Tok	Nitrofen
Tolkan	Isoproturon
Top Hand	Acetochlor
Topsin	Thiophanate-methyl
Tordon	Picloram
Tordon RTU	2,4-D
	Picloram
Torpedo	Permethrin
Totril	Ioxynil
Tre-Hold	NAA
Treflan	Trifluralin
Tri-Scept	Imazaquin
Tribunil	Methabenzthiazuron
Tributyltin	Tributyltin Chloride
Trigard	Cyromazine
Trooper	Dicamba
Tupersan	Siduron
Turbo	Metolachlor
	Metribuzin

TABLE A-2   *(Continued)*

Synonym or trade name	Pesticide(s) referred to by synonym or trade name
Turfcide	PCNB
Ureabor	Bromacil
Urox	Bromacil
Vapam	Metam-sodium
Velpar	Hexazinone
Vendex	Fenbutatin-oxide
Vernam	Vernolate
Vitavax-200	Carboxin
	Thiram
Vitavax-34	Carboxin
Vorlan	Vinclozolin
Vorlex	1,3-dichloropropene
	Methyl isothiocyanate
Vydate	Oxamyl
Weddar MCPA	MCPA
Weed-Out 6-Low Volatile Ester	2,4-D (butoxyethanol ester)
Weed-rhap LV 6D	2,4-D (butoxyethanol ester)
Weedar	2,4,5-T
Weedar 64	2,4-D
Weedar 64-A	2,4-D
Weedar Emulsamine	2,4-D
Weedmaster	2,4-D
	Dicamba
Weedone 170	2,4-D
	Dichlorprop
Weedone 2,4-DP	2,4-D
Weedone 638	2,4-D
Weedone CB	2,4-D
	Dichlorprop
Weedone LV-4	2,4-D (butoxyethanol ester)
Weedone LV4	2,4-D
Weedone LV6	2,4-D
Whip	Fenoxaprop-ethyl
XL	Benefin
	Oryzalin
Zolone	Phosalone
Zorial Rapid 80	Norflurazon

expect values to be identical. Generally speaking, the concentrations are based on the concept that there is an acceptable daily intake (ADI) for a pesticide and that a certain fraction (usually 20 percent) of one's exposure to the pesticide could come from drinking water. A safety factor of some kind is then applied, given the uncertainties in extrapolating from one test species (rats, mice, monkeys, etc.) to man. The safety factor also accounts for the range of sensitivities to chemicals typically exhibited by different individuals, even when exposed to similar levels of the chemical. In the case of the United States, both HALs and MCLs are given when they exist, with the HALs given in parentheses. The definitions of these two types of levels are given in Chapter 5, but the essential difference is that an MCL is a regulatory standard that public water supplies must meet, whereas the HAL only provides information about relative toxicity.

References
Water Quality Criteria
Australia Water Resource Council. 1987. *Standard for Maximum Residue Limits of Pesticide, Agricultural Chemical, Feed Additives, Veterinary Medicines and Noxious Substances in Food.* NH&MRL Guidelines for Drinking Water Quality in Australia. Canberra, Australia.

Canada Minister of National Health and Welfare. 1987. *Guidelines for Drinking Water Quality.* Federal-Provincial Advisory Committee on Environmental and Occupational Health. Ottawa, Canada.

Budesgesundhbl. 1989. *Empfehlung des Bundesgesundheitsamtes zum Vollzug der Trinwasserverorordung (TrinkwV) vom 22.* Mai 1986 (BGBl, I S. 760). Quelle: Bekanntmachungen des BGA. Berlin, Germany.

Healey, M.G., and A.H.H. Jones. 1989. Memorandum to the Chief Executives of Water Service Companies and Secretaries of Water Companies in England and Wales. Drinking Water Division, Department of Environment, London, England.

United States Environmental Protection Agency. 1990. Maximum Contaminant Levels, Federal Register 55:143. July 25, 1990.

United States Environmental Protection Agency. 1989. Office of Drinking Water Lifetime Health Advisories, Washington, D.C.

Physical Property Data for Pesticides
Bai, Q., Y. Zhai, S. Tian, Y. Wang, and H. Liu. 1988. Phosalone applied to cotton: its deposition and persistence on foliage, soils and final residues in seeds. *J. Environ. Sci. Health* B(21):33–43.

Basham, G., T.L. Lavy, L.R. Oliver, and H.D. Scott. 1987. Imazaquin persistence and mobility in three Arkansas soils. *Weed Science* 35:576–82.

Basham, G.W., and T.L. Lavy. 1987. Microbial and photolytic dissipation of imazaquin in soil. *Weed Science* 35:865–70.

Bilkert, J.N., and P.S.C. Rao. 1985. Sorption and leaching of three nonfumigant nematicides in soils. *J. Environ. Sci. Health* B20:1–26.

Blair, A.M., and T.D. Martin. 1988. A review of the activity, fate and mode of action of sulfonylurea herbicides. *Pestic. Sci.* 22:195–219.

Bowman, B.T. 1988. Mobility and persistence of metolachlor and aldicarb in field lysimeters. *J. Environ. Qual.* 17:689–94.

Dean, J.D., P.P. Jowise, and A.S. Donigian. 1984. *Leaching evaluation of agricultural chemicals (LEACH) Handbook.* Athens, Ga.: U.S. EPA. 407 pp.

Donigian, A.S., and R.F. Carsel. 1987. Modeling the impact of conservation tillage practices on pesticide concentrations in ground and surface waters. *Environ. Tox. Chem.* 6:241–50.

Ekler, Z. 1988. Behavior of thiocarbamate herbicides in soils: adsorption and volatilization. *Pestic. Sci.* 22:145–57.

Elabd, H., W.A. Jury, and M.M. Cliath. 1986. Spatial variability of pesticide adsorption parameters, *Environ. Sci. Technol.* 20:256–60.

Fadayomi, O., and G.F. Warren. 1977. Adsorption, desorption, and leaching of nitrofen and oxyfluorfen. *Weed Science* 25:97–100.

Gallandt, E.R., P.K. Fay, and W.P. Inskeep. 1989. Clomazone dissipation in two Montana soils. *Weed Tech.* 3:146–50.

Hamaker, J.M., 1975. Interpretation of soil leaching experiments. In *Environmental Dynamics of Pesticides,* ed. R. Haque and V.H. Freed. New York: Plenum. pp. 85–120.

Harris. 1988. A Comparison of the Persistence in a clay loam of single and repeated annual applications of seven granular insecticides. *J. Environ. Sci. Health* B23:1–23.

Johnson, B. 1988. *Setting Revised Specific Numerical Values: November 1988.* Sacramento: CDFA. 24 pp.

Jotcham, J.R., D.W. Smith, and G.R. Stephenson. 1989. Comparative persistence and mobility of pyridine and phenoxy herbicides in soil. *Weed Tech.* 3:155–61.

Jury, W.A., D.D. Focht, and W.J. Farmer. 1987. Evaluation of pesticide groundwater pollution potential from standard indices of soil-chemical adsorption and biodegradation. *J. Environ. Qual.* 16:422–28.

Kubiak, R., F. Fuhr, W. Mittelstaedt, M. Hansper, and W. Steffens. 1988. Transferability of lysimeter results to actual field situations. *Weed Science* 7:514–18.

Melancon, S.M., J.E. Pollard, and S.C. Hern. 1986. Evaluation of SESOIL, PRZM and PESTAN in a laboratory column leaching experiment. *Environ. Tox. Chem.* 5:865–78.

Mersie, W., and C.L. Foy. 1986. Adsorption, desorption, and mobility of chlorsulfuron in soils, *J. Agric. Food Chem.* 34:89–92.

Moncorge, J.M., and M.W. Murphy. 1987. The use of WL95481 in transplanted paddy rice. In British Crop Protection Conference. Surrey: BCPC. pp. 197–98.

Mueller, T.C., and P.A. Banks. 1990. Flurtamone adsorption and mobility in three Georgia soils. *Weed Science* 38:411–15.

Mueller, T.C., P.A. Banks, and D.C. Bridges. 1990. Dissipation of flurtamone in three Georgia soils. *Weed Science* 38:411–15.

Nash, R.G. 1988. *Dissipation from Soil, in Environmental Chemistry of Herbicides, Volume I,* ed. R. Grover. Boca Raton: CRC Press. pp. 131–70.

Nicholls, P.H. 1988. Factors influencing entry of pesticides into soil water. *Pestic. Sci.* 22:123–37.

Oyamada, M., and S. Kuwatsuka. 1987. Effects of soil properties and conditions on the degradation of three diphenyl ether herbicides in flooded soils. *J. Pesticide Sci.* 13:99–105.

Peter, C.J., and J.B. Weber. 1985. Adsorption and efficacy of trifluralin and butralin as influenced by soil properties. *Weed Science* 33:861–67.

Peter, C.J., and J.B. Weber. 1985. Adsorption, mobility and efficacy of alachlor and metolachlor as influenced by soil properties. *Weed Science* 33:874–81.

Peter, C.J., and J.B. Weber. 1985. Adsorption, mobility and efficacy of alachlor and metolachlor as influenced by soil properties. *Weed Science* 33:874–81.

Rao, P.S.C., A.G. Hornsby, and R.E. Jessup. 1985. Indices for ranking the potential for pesticide contamination of groundwater. In Proceedings of the 44th Annual Meeting of the Soil and Crop Science Society of Florida. Jacksonville: SCSSF. pp. 1–8.

Roy, D.N., S.K. Konar, D.A. Charles, J.C. Feng, R. Prasad, and R.A. Campbell. 1989. Persistence, movement, and degradation of hexazinone in selected Canadian boreal forest soils. *J. Agric. Food Chem.* 37:437–40.

Royal Society of Chemistry. 1987. *The Agrochemicals Handbook.* Nottingham: RSC. 884 pp.

Rueppel, M.L., B.B. Brightwell, J. Schaefer, and J.T. Marvel. 1977. Metabolism and degradation of glyphosate in soil and water. *J. Agric. Food Chem.* 25:517–28.

Sabljic, A. 1987. On the prediction of soil sorption coefficients of organic pollutants from molecular structure: application of molecular topology model. *Environ. Sci. Technol.* 21:358–66.

Sanchez-Camazano, M., and M.J. Sanchez-Martin. 1988. Influence of soil characteristics on the adsorption of pirimicarb. *Environ. Tox. Chem.* 7:559–64.

Unger, M.A., W.G. Macintyre, and R.J. Huggett. 1988. Sorption behavior of tributyltin on estuarine and freshwater sediments. *Environ. Toxic. Chem.* 7:907–15.

U.S. EPA. 1988. *Health Advisories for 50 Pesticides,* Pb88-245931. 861 pp.

Walker, A., and P.A. Brown. 1985. The relative persistence in soil of five acetanilide herbicides. *Bull. Environ. Contam. Toxicol.* 34:143–49.

Wauchope, R.D., T.M. Buttler, A.G. Hornsby, P.W.M. Augustijn- Beckers, and J.P. Bert. 1992. The SCS/ARS/CES pesticide properties database for environmental decision-making. *Reviews of Environmental Contamination and Toxicology,* 123:1–164.

Weed Science Society of America. 1983. *Herbicide Handbook,* Fifth Edition, Champaign: WSSA. 515 pp.

Wehtje, G., R. Dickens, J.W. Wilcut, and B.F. Hajek. 1987. Sorption and mobility of sulfometuron and imazapyr in five Alabama soils. *Weed Science* 35:858–64.

Wilkerson, M.R., and K.D. Kim. 1986. The pesticide contamination prevention act: setting specific numerical values. Sacramento: CDFA. 27 pp.

Wood, L.S., H.D. Scott, D.B. Marx, and T.L. Lavy. 1987. Variability in sorption coefficients of metolachlor on a captina silt loam. *J. Environ. Qual.* 16:251–56.

Wu, T.L. 1980. Fate and transport of alachlor and atrazine in a Maryland watershed. *J. Environ. Qual.* 9:459–65.

Zhuang, W., C.S. Helling, T.J. Gish, C.B. Coffman, A.R. Isensee, and P.C. Kearney. 1985. Persistence, leaching and groundwater residues of atrazine, alachlor, and cyanazine under no-tillage practices. Submitted to *J. Agric. Food Chem.*

Zimdahl, R.J., and S.K. Clark. 1982. Degradation of three acetanilide herbicides in soil. *Weed Science* 30:545–48.

Appendix II

Additional Information on Models

PRZM/RUSTIC

The PRZM/RUSTIC model is a FORTRAN computer program that was developed on a VAX minicomputer. A PC version of the original PRZM model is available, and a PC version of the recently updated PRZM model (including volatilization and metabolites) has more recently been made available (see the end of this appendix for ordering information).

The PRZM portion of the RUSTIC model is taken from the earlier stand-alone program. PRZM is a one-dimensional, dynamic, compartmental model that can be used to simulate chemical movement in unsaturated soil systems within and immediately below the plant root zone. It has two major components: water and chemical transport. The hydrologic component for calculating runoff and erosion is based on the Soil Conservation Service curve number technique and the Universal Soil Loss Equation. Evapotranspiration is estimated either directly from pan evaporation data, or based on an empirical formula. Evapotranspiration is divided among evaporation from crop interception, evaporation from soil, and transpiration by the crop. Water movement is simulated by the use of generalized soil parameters, including field capacity, wilting point, and saturation water content. With a newly added feature, irrigation may also be considered.

The chemical transport component can simulate pesticide application on the soil or on the plant foliage. Dissolved, sorbed, and vapor-phase concentrations in the soil are estimated by simultaneously considering the processes of pesticide uptake by plants, surface runoff, erosion, decay, volatilization, foliar washoff, advection, dispersion, and retardation. Two options are now available to solve the transport equation:

1. The original backwards-difference implicit scheme, which may be affected by excessive numerical dispersion at high Peclet numbers; or
2. The "method of characteristics" algorithm, which eliminates numerical dispersion while increasing model execution time.

Predictions are made on a daily basis. Output can be summarized for a daily, monthly, or annual period. Daily time series values of various fluxes or storages can be written to sequential files during program execution for subsequent analysis. In addition, a "Special Actions" option allows the user to output soil profile pesticide concentrations at user-specified times. It is also possible to change the values of certain parameters during the simulation period.

Several limitations in the PRZM program were obvious in the original 1984 release of the program. These fell into four basic categories:

1. Hydrology;
2. Soil hydraulics;
3. Method of solution of the transport equation; and
4. Deterministic nature of the model.

Hydrology and soil hydraulic computations are still performed, using a daily time step, even though some of the processes involved (evaporation, runoff, erosion) would obviously require a finer time step to improve accuracy.

For instance, simulation of erosion by runoff depends on the peak runoff rate, which is in turn dependent upon the time base of the runoff hydrograph. This depends to some extent upon the duration of the precipitation event. PRZM retains its daily time step due primarily to the relative unavailability of precipitation data in anything but a daily format. This limitation has in part been mitigated by enhanced parameter guidance.

In PRZM 1.0, the soil hydraulics were simple—all drainage to field capacity water content was assumed to occur within 1 day. (An option to make drainage time dependent was included, but there is little evidence that this option was ever used.) This had the effect, especially in deeper soils, of inducing overly rapid movement of water and therefore pesticide through the soil profile. While this representation of soil hydraulics has been retained in PRZM 2.0, the user can couple PRZM to VADOFT in order to simulate behavior in the vadose zone more accurately. VADOFT utilizes a Galerkin finite element technique to approximate the Richard's Equation governing unsaturated flow, and is thus a much more rigorous technique for handling soil hydraulics.

The addition of volatilization to PRZM 2.0 has brought into focus another limitation of the soil hydraulics representation. PRZM simulates only downward advective movement of draining water and does not account for transport of pesticide in response to the diffusive transport of water, such as when evaporation upward or wicking pulls water back up toward the soil surface. This process has been identified by Jury and other researchers as a relatively important one when trying to simulate the volatilization of pesticides from soil.

Another limitation of PRZM 1.0 was the inadequacy of the method of solution (backwards difference) for the transport equation. In advection dominated systems at high Peclet number, this solution method leads to a high degree of numerical dispersion or smearing of the concentration profiles. Since much of the amount of chemical that exits the root zone is usually determined by how fast the leading edge of the plume gets below the root zone, numerical dispersion can result in severe overestimation of the quantity of material leaching into groundwater. In PRZM 2.0, a new formulation (Method of Characteristics) is available for solving the transport equation. This approach effectively eliminates numerical dispersion, but it does force the user to put an accurate value for the dispersion coefficient into the model, unlike PRZM 1.0, which simulated extensive dispersion whether the user wanted it to occur or not.

The final limitation is the use of field-average water and chemical transport parameters to represent a spatially heterogeneous matrix (that is, soil). Several researchers have shown that this approach leads to slower breakthrough times than are observed using stochastic approaches. This concern has been addressed by adding the capability to run PRZM in a Monte Carlo framework. Thus, distributional rather than field-average values can be utilized as input to the model, which will then produce distributional outputs of the relevant parameters, such as chemical flux to the water table.

VADOFT (Vadose Zone Flow and Transport Model) is a finite-element code for simulating moisture movement and solute transport in the vadose zone. It is the second part of the three-component RUSTIC model for predicting the movement of pesticides within and below the plant root zone and assessing subsequent groundwater contamination. The VADOFT code simulates one-dimensional, single-phase moisture and solute transport in unconfined, variably saturated porous media. Transport processes include hydrodynamic dispersion, advection, linear equilibrium sorption, and first-order decay. The code predicts infiltration of recharge rate and solute mass flux entering the saturated zone. Parent/daughter chemical relationships may be simulated.

SAFTMOD (Saturated Zone Flow and Transport Model) is a finite-element code for simulating groundwater flow and transport in the saturated zone. Since support for it within the RUSTIC model has recently been dropped, it will receive no further attention here.

The linkage of PRZM 2.0, VADOFT, and SAFTMOD presented a significant technical challenge to the authors of RUSTIC. The primary problems had to deal with the issues of time step synchronization and spatial linkages. The daily time step is maintained within PRZM, but VADOFT sometimes needs a time step of only several minutes in order to obtain convergence with the highly nonlinear Richard's equation. The challenges associated with linkage to SAFTMOD were among the many reasons that support for SAFTMOD was dropped.

EXAMS

EXAMS is the standard model for predicting pesticide behavior in small ponds. It is a computer-based system for installing and running chemical simulation studies with models of aquatic ecosystems. EXAMS's environmental models are maintained in a file composed of concise ("canonical") descriptions of aquatic systems. Each water-body is represented via a set of segments or distinct zones in the system. The program is based on a series of mass balances for the segments that give rise to a single differential equation for each. Working from the transport and transformation process equations, EXAMS compiles an equation for the net rate of change of chemical concentration in each segment. The resulting system of N differential equations describes the mass balance for the entire system. EXAMS includes a descriptor language that simplifies the specification of system geometry and connectedness. The code is written in a general (N-segment) form. The software is available in 32- (MS-DOS) and 50- (VAX) segment versions.

The second-order process models used to compute the kinetics of chemicals are the central core of EXAMS. Each includes a direct statement of the interactions between the chemistry of a compound and the environmental forces that shape its behavior in aquatic systems. Most of the process equations are based on standard theoretical constructs or accepted empirical relationships. For example, the light intensity in the water column of the system is computed using the Beer-Lambert law, and temperature corrections for rate constants are computed using Arrhenius functions. Ionization of organic acids and bases, complexation with dissolved organic matter (DOC), and sorption of the compound with sediments and biota are treated as thermodynamic properties or (local) equilibria that modify the speed of the kinetic processes. For example, an organic base in the water column may occur in a number of molecular species (as dissolved ions, sorbed with sediments, etc.), but only the uncharged, dissolved species can be volatilized across the air-water interface. EXAMS allows for the simultaneous treatment of up to 28 molecular species of a chemical—the parent uncharged molecule, and singly, doubly, or triply charged cations and anions, each of which can occur in a dissolved, DOC-complexed, sediment-sorbed, or biosorbed form. The program computes the fraction of the total concentration of a compound that is present as each of the 28 molecular structures ("distribution coefficients," ALPHA). These values enter the kinetic equations as multipliers on the rate constants. The program thus completely accounts for differences in reactivity that depend on the molecular form of the chemical. EXAMS makes no internal assumptions about the relative transformation reactivities of the 28 molecular species. These assumptions are under direct user control through the way the user structures the chemical input data.

EXAMS includes two algorithms for computing the rate of photolytic transformation of a synthetic organic chemical. These algorithms accommodate the two more common kinds of laboratory data and chemical parameters used to describe photolysis reactions. The simpler algorithm requires only an average pseudo–first-order rate constant (KDP) applicable to near-surface waters under cloudless conditions at a specified reference latitude (RFLAT). In order to give the user control of reactivity assumptions, KDP is coupled to user-supplied (normally unit-valued) reaction quantum yields (QUANT) for each molecular species of the compound. This approach makes possible a first approximation of photochemical reactivity, but neglects the very important effects of changes in the spectral quality of sunlight with increasing depth in a water body. The more complex photochemical algorithm computes photolysis rates directly from the absorption spectra (molar extinction coefficients) of the compound and its ions, measured values of the reaction quantum yields, and the environmental concentrations of competing light absorbers (chlorophylls, suspended sediments, dissolved organic carbon, and water itself).

The total rate of hydrolytic transformation of a chemical is computed by EXAMS as the sum of three contributing processes. Each of these processes can be entered via simple rate constants or as Arrhenius functions of temperature. The rate of specific-acid catalyzed reactions is computed from the pH of each sector of the ecosystem, and specific-base catalysis is computed from the environmental pOH data. The rate data for neutral hydrolysis of the compound is entered as a set of pseudo-first-order rate coefficients (or Arrhenius functions) for reaction of the 28 (potential) molecular species with the water molecule.

EXAMS allows the user to compute biotransformation of the chemical in the water column and, in the bottom sediments, of the system as entirely separate functions. Both functions are second-order equations that relate the rate of biotransformation to the size of the bacterial population actively degrading the compound. The second-order rate constants (KBACW for the water column, KBACS for benthic sediments) can be entered either as single-valued constants or as functions of temperature. When a nonzero value is entered for the Q-10 of a biotransformation (parameters QTBAW and QTBAS, respectively), KBAC is interpreted as the rate constant at 20°C, and the biolysis rate in each sector of the ecosystem is adjusted for the local temperature (TCEL).

Oxidation reactions are computed from the chemical input data and the total environmental concentrations of reactive oxidizing species (alkylperoxy and alkoxyl radicals, etc.) specified by the user, corrected for ultraviolet light extinction in the water body. The chemical data can again be entered either as simple second-order rate constants or as Arrhenius functions. Oxidations due to singlet oxygen are computed from chemical reactivity data and singlet oxygen concentrations estimated as a function of dissolved organic carbon (DOC) concentration, oxygen tension, and light intensity. Reduction is included in the program as a simple second-order reaction process driven by the user entries for concentrations of reductants in the system.

Internal transport and export of a chemical occur in EXAMS via advective and dispersive movement of dissolved, complexed, sediment-sorbed, and biosorbed materials, and by volatilization losses at the air-water interface. EXAMS provides a set of vectors (JFRAD, et al.) that allows the user to specify the location and strength of both advective and dispersive transport pathways. Advection of water through the system is then computed from the water balance, using hydrologic data (rainfall, evaporation rates, stream flows, groundwater seepages, etc.) supplied as part of the definition of each environment. Dispersive interchanges within the system and across system boundaries are computed from the characteristic length (CHARL), cross-sectional area (XSTUR), and dispersion coefficient (DSP) specified for each active exchange pathway. EXAMS can compute transport of a chemical via whole-sediment bedloads, suspended sediment washloads, groundwater infiltration, transport through the thermocline of a lake, losses in effluent streams, and so forth. Volatilization losses are computed using a two-resistance model. This computation treats the total resistance to transport across the air-water interface as the sum of resistances in the liquid and vapor phases immediately adjacent to the interface.

EXAMS allows for entry of external loadings of chemicals via point sources, nonpoint sources, dry fallout or aerial drift, atmospheric washout, and groundwater seepage entering the system. Any type of chemical load can be entered for any system segment, but the program will not implement a loading that is inconsistent with the system definition. For example, the program will automatically cancel a rainfall load entered for the hypolimnion or benthic sediments of a lake ecosystem. When this type of corrective action is executed, the change is reported to the user via an error message.

The user can select among three operating "modes" to choose the level of complexity needed to address the problem being studied. In the simplest case (mode 1), EXAMS executes a direct steady-state solution of the dynamic system equations, thus generating a long-term analysis using a single set of environmental conditions (for example, annual average driving forces). In mode 2, EXAMS makes available initial-value approaches that can be used to set initial conditions and introduce immediate "pulse" chemical loadings. To the extent that changes in hydrographic volumes (for example, during spates) can be neglected, this mode can be used to evaluate shorter-term fate and transport events by segmenting the input datasets and simulation intervals according to time slices under full user control. In mode 3, EXAMS uses a set of 12 monthly values of all environmental parameters, with input loads that can change monthly and can also include pulse events on individual dates, to compute the dynamics of chemical contamination over the course of 1 or more years' time. The outputs produced by the system are analogous for all modes of operation, although they differ in detail. For example, in mode 1, a summary table and sensitivity analyses of system fluxes are reported for steady-state conditions; in mode 2, the reports are generated for conditions at the close of each time slice; and in mode 3, the program reports annual (or interannual) average values and the size and location of exposure extrema.

EXAMS has been designed to evaluate the consequences of longer-term, primarily time-averaged chemical loadings that ultimately result in trace-level contamination of aquatic systems. EXAMS generates a steady-state, average flow field (long-term or monthly) for the ecosystem. The program thus cannot fully evaluate the transient concentrated EECs that arise (for example, from chemical spills). This limitation derives from two factors. First, a steady flow field is not always appropriate for evaluating the spread and decay of a major pulse (spill) input. Second, an assumption of trace-level EECs, which can be violated by spills, has been used to design the process equations used in EXAMS. The following assumptions were used to build the program.

Firstly, it is assumed that a useful evaluation can be executed independently of the chemical's actual effects on the system. In other words, it is assumed that the chemical itself will not radically change the environmental variables that drive its transformations. Thus, for example, it is assumed that an organic acid or base will not change the pH of the system; it is assumed that the compound itself will not absorb a significant fraction of the light entering the system; bacterial populations do not significantly increase (or decline) in response to the presence of the chemical.

Secondly, EXAMS uses linear sorption isotherms and second-order (rather than Michaelis-Menten or Monod) expressions for biotransformation kinetics. This approach is known to be valid for low concentrations of pollutants; its validity at high concentrations is less certain. EXAMS controls its computational range to ensure that the assumption of trace-level concentrations is not grossly violated. This control is keyed to aqueous-phase (dissolved) residual concentrations of the compound: EXAMS aborts any analysis generating EECs that exceed (the lesser of) 50 percent of the compound's aqueous solubility or 10 micromolar concentrations of a unionized-ionized molecular species. This restraint incidentally allows the program to ignore precipitation of the compound from solution, and precludes inputs of solid particles of the chemical.

Sorption is treated as a thermodynamic or constitutive property of each segment of the system (that is, sorption/desorption kinetics are assumed to be rapid compared to other processes). The adequacy of this assumption is partially controlled by properties of the chemical and system being evaluated. Extensively sorbed chemicals tend to be sorbed and desorbed more slowly than weakly sorbed compounds; desorption half-lives may approach 40 days for the most extensively bound compounds. Experience with the program has indicated, however, that strongly sorbed chemicals tend to be captured by benthic sediments, where their release to the water column is controlled by benthic exchange processes. This phenomenon overwhelms any accentuation of the speed of processes in the water column that may be caused by the assumption of local equilibrium.

Input parameters include:

1. A set of chemical loadings on each sector of the ecosystem;
2. Molecular weight, solubility, and ionization constants of the compound;
3. Sediment-sorption and biosorption parameters, biomasses, benthic water contents and bulk densities, suspended sediment concentrations, sediment organic carbon, and ion exchange capacities;
4. Volatilization parameters, Henry's Law constant or vapor pressure data, wind speeds, and reaeration rates;
5. Photolysis parameters, reaction quantum yields, absorption spectra, stratospheric ozone, cloudiness, relative humidity, atmospheric dust content and air-mass type, scattering parameters, suspended sediments, chlorophyll, and dissolved organic carbon;
6. Hydrolysis, second-order rate constants or Arrhenius functions for the relevant molecular species; pH, pOH, and temperatures;
7. Oxidation, rate constants, temperatures, surface oxidant concentrations, dissolved organic carbon, and oxygen tension;
8. Biotransformation rate constants, temperatures, bacterial population densities;
9. Parameters defining strength and direction of advective and dispersive transport pathways; and

10. System geometry and hydrology, volumes, areas, depths, rainfall, evaporation rates, entering stream and nonpoint-source flows and sediment
loads, and groundwater flows.

Although EXAMS allows for the entry of extensive environmental data,
the program can be run with a much reduced data set when the chemistry of
a compound of interest precludes some of the transformation processes. For
example, pH and pOH data can be omitted in the case of neutral organics
that are not subject to acid or alkaline hydrolysis. The 20 output tables include
an echo of the input data, as well as tabulations giving the exposure, fate, and
persistence of the chemical. The program prints a summary report of the
results obtained. Printer-plots of longitudinal and vertical concentration
profiles, as well as time-based graphics, can be invoked by the user.

To Order Either the PRZM/RUSTIC or EXAMS Model

Contact:
Model Distribution Coordinator
Center for Exposure Assessment Modeling
U.S. Environmental Protection Agency
960 College Station Road
Athens, GA 30605-2720
phone: 706/546-3549
fax: 706/546-3126

Glossary

Aquifer   a layer of rock, sand, or gravel that will yield usable supplies of water when pumped

Aquifer (artesian)   an aquifer in which groundwater is held under pressure by the presence of confining layers, causing water to rise above the top of the aquifer for any wells placed in it

Aquifer (principal)   the aquifer in a given area that is the most important as a source of drinking and irrigation

Aquifer (secondary)   any aquifer in a region other than the principal aquifer

Aquifer (sole source)   as designated under the Safe Drinking Water Act (SDWA), the principal aquifer in a region where no secondary aquifers or usable surface water sources are present

Aquifer (surficial)   an aquifer that has no confining layer above it, such that the water at its top is under atmospheric pressure (also known as a phreatic aquifer)

Cone of Depression   the depression in the water table around a well created by the pumped withdrawal of water

Contaminant   any physical, chemical, biological, or radiological substance or material in water

Discharge   the flow of surface water in a stream or canal or the outflow of groundwater from a flowing artesian well, ditch, or spring

Drainage Well   a well drilled to carry excess water down and off of agricultural fields

Drawdown   the vertical drop of the water level in a well during pumping operations

Groundwater   water below the land surface that is at or above atmospheric pressure

Hardness   a characteristic of water caused by the presence of various salts of calcium, magnesium, and iron

Health Advisory Level (HAL)   a nonregulatory, health-based reference level of pesticide concentrations in drinking water at which there are no adverse health effects when ingested over various periods of time (1 day, 10 days, lifetime)

231

Leaching   the downward movement with water by dissolved or suspended minerals, fertilizers, pesticides, or other solutes through the soil

Maximum Contaminant Level (MCL)   an enforceable regulatory standard for the maximum permissible concentration of a pesticide or other contaminant in drinking water, required to be set as close to the relevant health advisory level (HAL) as is technically feasible

Mining an Aquifer   the withdrawal of groundwater from an aquifer that exceeds the rate of recharge of the aquifer

Nitrate   an important plant nutrient and type of inorganic fertilizer resulting from septic systems, animal feed lots, agricultural fertilizers, manure, and certain nitrogen-fixing plants such as legumes

Nonpoint Source   a widespread source of drinking water contaminants, such as a large field crop treated with pesticide

No Effect Level (NOEL)   the exposure level, generally expressed as mg/kg/day, at which no adverse effects are noted in laboratory toxicity studies of a pesticide or other toxicant on a test species such as a rat or mouse also denoted the NOAEL

Perched Water Table   a zone of unpressurized water held above a lower aquifer by a layer of relatively impermeable material

Permeability   the capacity of a porous rock, sediment, or soil to transmit water

Parts Per Billion (PPB)   a commonly used unit of concentration regarding the occurrence of pesticides in drinking water supplies; a proportion expressing the parts of pesticide dissolved in 1,000,000,000 parts of water, numerically close to μg/L or micrograms per liter, a more scientifically sound concentration

pH   a numerical measure of water acidity equal to the negative of the common logarithm of the hydrogen ion concentration in the water (values less than 7 are acidic, and values greater than 7 are basic)

Point Source   a single, relatively small, easily distinguishable and demonstrable source of a drinking water contaminant

Porosity   the relative volume of a porous medium such as sand, gravel, rock, or soil that may be occupied by a permeating fluid or gas such as water or air

Recharge Zone   an area of land that is sufficiently permeable to allow water to penetrate down into a region's aquifers

Salinity   the concentration of dissolved salts, especially oceanic-derived materials, in water

Saturated Zone   a portion of the soil profile where all pores are filled with water and the water is at or above atmospheric pressure

Turbidity   a measure of water cloudiness caused by the occurrence of suspended solids

Unsaturated Zone   a portion of the soil profile that contains both water and air and within which the water is never at pressures above atmospheric

Water Table   the top of the uppermost aquifer in a region, below which all pores are generally filled with water

Watershed   all land and water within a drainage unit

Index